APPLICATIONS OF ADVANCED STRAIN MEASUREMENT TECHNIQUES

110

BRITISH SOCIETY FOR STRAIN MEASUREMENT

Whittles Publishing
Caithness

Published by
Whittles Publishing Services, Roseleigh House,
Harbour Road, Latheronwheel, Caithness KW5 6DW

British Library Cataloguing in Publication Data

Advanced measurement techniques (*Conference: 1987: London, England*)
Application of advanced strain measurement techniques.
1. Materials. Strains. Measurement
I. British Society for Strain Measurement
620.1'123
ISBN 1-870325-05-2

Printed in Great Britain by
Bell and Bain Ltd, Scotland

CONTENTS

SECTION I COMPOSITE MATERIALS

SECTION II DYNAMICS AND STRUCTURES

SECTION III FRACTURE MECHANICS

FOREWORD

The measurement of strain is of great importance to those concerned with the design of engineering structures and mechanisms, material investigation and testing, product reliability and integrity, and many other aspects of design, development, manufacture and use of engineering products.

The increasing use of composite materials in engineering, the behaviour of components under high dynamic loads and the importance of the study of fracture all demand refinements to existing measurement techniques or the introduction of new methods of measurement. An awareness of these improvements is thus of extreme importance to all who are engaged in the study of engineering materials in its broadest sense.

With this in mind, an international conference was held in London in August 1987 under the auspices of the British Society for Strain Measurement and the Society for Experimental Mechanics (USA), and this conference attracted many very interesting papers presented by internationally recognised experts.

Following the success of this conference, the British Society for Strain Measurement felt that in view of its quality and importance, the subject matter should be made available to a wider audience. This book is the result.

It will, I hope, find a place on the bookshelves of technical libraries, laboratories engaged in research and development, quality control and materials testing laboratories and the personal libraries of those working in the field of strain measurement. There is also much of interest for those working in related disciplines. As an example, there are a number of papers on optical methods of measurement which might have applications to related fields of enquiry.

S. J. COLLINS
Chairman
British Society for Strain Measurement

CONTRIBUTORS

M. K. Abdelhamid	Department of Engineering Science and Mechanics, Iowa State University, USA.
E. R. Akum	Department of Mechanical Engineering, Imperial College of Science and Technology, UK.
D. W. Begg	Department of Civil Engineering, Portsmouth Polytechnic, UK.
D. J. Berry	Ometron Ltd., Worsley Bridge Road, London, UK.
M. Bourke	Department of Mechanical Engineering, Imperial College of Science and Technology, UK.
P. Brooks-Johnson	Department of Mechanical Engineering, Imperial College of Science and Technology, UK.
J. E. Butler	Department of Civil Engineering, Portsmouth Polytechnic, UK.
B. W. Chang	School of Engineering and Applied Science, University of California, Los Angeles, CA, USA.
G. L. Cloud	Department of Metallurgy, Mechanics and Materials Science, Michigan State University, East Lansing, MI 48823, USA.
W. T. Evans	Department of Mechanical and Production Engineering, The Polytechnic of Wales, Pontypridd, UK.
G. M. Everett	Ometron Ltd., Worsley Bridge Road, London, UK.
M. E. Fourney	School of Engineering and Applied Science, University of California, Los Angeles, CA, USA.
J. A. Gilbert	Department of Mechanical Engineering, University of Alabama in Huntsville, Huntsville, Alabama 35899, USA.
V. Grabulov	Belgrade Institute of Technology, Yugoslavia.
P. Greguss	Applied Biophysics Laboratory, Technical University, Budapest, Hungary.

V. Grubisic	Fraunhofer-Institut fur Betriebsfestigkeit (LBF), Darmstadt, FRG.
E. J. Hearn	Department of Mechanical and Production Engineering, City of Birmingham Polytechnic, Birmingham, UK.
A. W. Hendry	Department of Civil Engineering and Building Science, University of Edinburgh, UK.
P. J. Herrera Franco	Instituto Technologico de Merida, Merida, Mexico.
J. Der Hovanesian	Department of Engineering, Oakland University, Rochester, MI 48063, USA.
Y. H. Hung	Department of Engineering, Oakland University, Rochester, MI 48063, USA.
D. L. Lehner	Department of Mechanical Engineering, University of Alabama in Huntsville, Huntsville, Alabama 35899, USA.
J. L. Lindner	Department of Mechanical Engineering, University of Alabama in Huntsville, Huntsville, Alabama 35899, USA.
A. S. Kobayashi	Department of Mechanical Engineering, University of Washington, Seattle, Washington 98195, USA.
K. G. McConnell	Department of Engineering Science and Mechanics, Iowa State University, USA.
H. J. MacGillivray	Department of Mechanical Engineering, Imperial College of Science and Technology, UK.
J. McKelvie	Department of Mechanical and Process Engineering, University of Strathclyde, Glasgow, UK.
P. M. MacKenzie	Department of Mechanical and Process Engineering, University of Strathclyde, Glasgow, UK.
J. Neugebauer	Fraunhofer-Institut fur Betriebsfestigkeit (LBF), Darmstadt, FRG.
G. Nicoletto	Department of Mechanical, Nuclear and Aeronautical Constructions and Metallurgy, University of Bologna, Italy.

R. H. Owens	Ometron Ltd., Worsley Bridge Road, London, UK.
T. M. E. Refai	Department of Civil Engineering, Kansas State University, Manhattan, Kansas, USA.
M. Rezvani	Department of Engineering Science and Mechanics, Virginia Polytechnic Institute and State University, Blacksburg, Virginia 24061, USA.
R. Royles	Department of Civil Engineering and Building Science, University of Edinburgh, UK.
R. M. Russell	Deltacam Systems, Birmingham, UK.
W. N. Sharpe Jnr	Department of Mechanical Engineering, The Johns Hopkins University, Baltimore, MD 21218, USA.
C. W. Smith	Department of Engineering Science and Mechanics, Virginia Polytechnic Institute and State University, Blacksburg, Virginia 24061, USA.
W. A. Stange	Aero Propulsion Laboratory, AFWAL/POTA, Wright-Patterson AFB OH 45433-6563, USA.
S. E. Swartz	Department of Civil Engineering, Kansas State University, Manhattan, Kansas, USA.
C. A. Walker	Department of Mechanical and Process Engineering, University of Strathclyde, Glasgow, UK.
N. Walker	Department of Mechanical and Production Engineering, City of Birmingham Polytechnic, Birmingham, UK.
G. A. Webster	Department of Mechanical Engineering, Imperial College of Science and Technology, UK.
P. J. Webster	Department of Civil Engineering, University of Salford, UK
K. Yang	Department of Mechanical Engineering, University of Washington, Seattle, Washington 98195, USA.
K. Zimmerman	Department of Metallurgy, Mechanics and Materials Science, Michigan State University, East Lansing, MI 48823, USA.

1 A Moiré interferometric hole drilling method for residual stress measurements

G.Nicoletto

SUMMARY

An alternative and semi-destructive method for measuring residual stresses near the surface of a component is presented. It is based on the use of a hole drilling procedure to locally relax the residual stresses in the material and the use of diffraction gratings to sense the resulting deformation. In-plane displacement components around the hole are subsequently revealed by Moiré interferometry. The procedure for the application of the method is described and the original design of a compact, portable interferometer is presented. The theoretical relationships yielding stresses as functions of displacement components are given and their accuracy checked against known loading cases.

INTRODUCTION

The ubiquitous presence of residual stresses in materials and components is well known to the design engineer [1]. Control of residual stress levels is an important requirement in structural design since premature failures in various industries are attributed to unfavourable interactions between service-induced cyclic stresses and residual stresses, especially in corrosive environments.

Several experimental techniques for residual stress measurement are presently available. They are usually categorized as nondestructive, semidestructive and destructive on the basis of the damage done to the component by their application. The most widely used semidestructive method is probably the hole-drilling strain-gauge method. In this method a small hole is drilled through a strain gauge rosette cemented on the component and the residual stresses are computed within the linear elasticity framework from the three strain readings. Since the original proposal of the method in 1950 [2], numerous authors have contributed to the present stage of development of this method, which is testified by a recently issued ASTM (American Society for Testing and Materials) standard practice [3].

The method for residual stress measurement to be presented hereafter is similar to the hole-drilling strain gauge method as it uses the drilling procedure to locally relax the residual stresses of the material. It differs in that a diffraction grating solidly attached to the surface replaces the strain gauge rosette. In-plane displacement field components are subsequently revealed using Moiré interferometry, a high sensitivity measuring technique based on physical optics concepts (i.e. diffraction and interference) using a coherent light source.

In this chapter, after a brief outline of Moiré interferometry, the procedure for specimen preparation, a specially designed, portable instrument and the theoretical stress vs. in-plane displacement relationships of the Moiré-interferometric hole-drilling method are described. Its initial applications to known stress systems are presented. The promising aspects of the method are discussed with reference to the present capabilities of the strain gauge based method.

BASICS OF MOIRÉ INTERFEROMETRY

Moiré interferometry is a full-field technique which reveals in-plane displacement components with high sensitivity. It represents a relatively new development in experimental mechanics [4]. In its basic configuration, schematically shown in Figure 1.1, a Moiré interferometric system is given by a collimated beam of coherent light partially folded by a mirror to provide symmetrical illumination of a linear, reflective diffraction grating attached onto the surface under investigation.

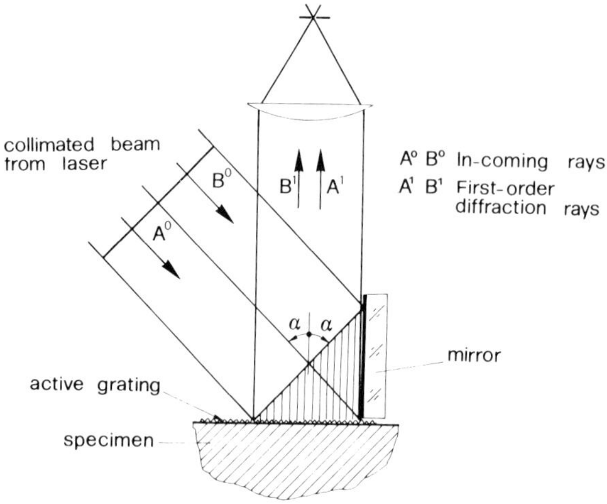

Figure 1.1

The phenomenon of two-beam interference in the volume of ray intersection defines the reference frequency f of the Moiré system which is given by

$$f = 2 \sin \alpha / \lambda \tag{1.1}$$

where λ is the wavelength of the light and α is the angle of incidence between in-coming rays and the normal to the surface. While α in equation (1.1) does not change during an experiment, the specimen grating frequency (initially equal to f/2) varies according to the geometric changes of the surface. The interaction of specimen grating and reference grating yields a moire pattern depicting the relative in-plane displacement component in the direction perpendicular to the grating lines. Fringe patterns are readily analyzed using the following equation [5],

$$U = gN = N/f \tag{1.2}$$

where U is the relative displacement component, N is the fringe order and g and f are the reference grating pitch and frequency, respectively. The displacement measuring sensitivity is in the sub-micron range since high frequency diffraction grating are used (i.e. f = 2400 1/mm implies a sensitivity of 0.417 micron per integer fringe order). From displacements, full-field strain distributions can be obtained by numerical, optical or mechanical differentiation. Local strain estimates can alternatively be determined from fringe gradient evaluation [4].

THE MOIRÉ INTERFEROMETRIC HOLE-DRILLING METHOD

The procedure of the Moiré-interferometric hole-drilling method is given by: (a) component preparation; (b) fringe visualization; (c) residual stress computation.

Component preparation

The procedure for component preparation consists of surface preparation, diffraction grating replica and hole-drilling. The replica of a diffraction grating on the specimen surface as well as the necessary precautions do not differ appreciably from a strain gauge cementing procedure. In fact surface preparation includes a thorough cleaning and degreasing while abrading and grinding should be limited to avoid generation of spurious residual stresses. In the specimen grating production shown schematically in Figure 1.2, a mould is initially positioned and pressed on liquid epoxy adhesive. A separate procedure for mould production already described in [4], is omitted here. After polymerization, the mould is pried off leaving a thin, reflective replica on the specimen surface. Since the present replica procedure makes use of flat, rigid moulds, it is restricted to nearly flat specimen surfaces. Although grating frequencies up to 4000 lines /mm have been successfully tested [4], a reference frequency

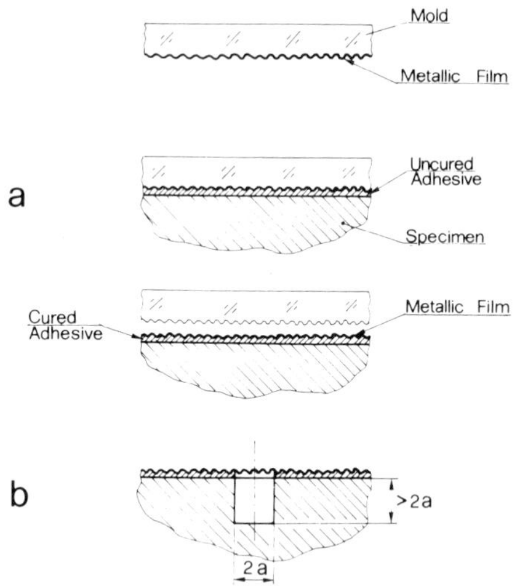

Figure 1.2

of 2400 lines/mm was selected for this study. Following the rosette approach, gratings characterized by three sets of linear furrows inclined at 0°, 45° and 90° were used.

Extensive studies with the hole-drilling strain gauge method have demonstrated the relevance of the drilling operation on subsequent residual stress measurements and several alternate drilling techniques to the simple end milling, such as abrasive jet machining and very high speed drilling have been tested [7]. In the initial applications of the present method, however, holes were drilled through diffraction gratings and into steel specimens only with end mills.

Fringe visualization

While the previous procedure of surface preparation for the Moiré interferometric hole drilling method is relatively simple, the low cost, straightforward applicability and capability of measurements in the field of the strain gauge technology are advantages apparently out of reach of the coherent optics methods of strain analysis. They are usually characterized by expensive and burdensome equipment, practical complexities and restriction to a vibration-free environment. In order to develop a method somewhat comparable to the strain gauge method, the design of the instrumentation for fringe visualization and recording after the hole drilling operation was aimed at achieving low cost and portability.

The cost of a coherent optics instrumentation is given by the cost of the laser and of the optical components. Since the replica procedure outlined above yields reflective gratings of high diffraction efficiency, laser power requirements are reduced when analyzing limited areas as in the present case. Small-power lasers, being increasingly inexpensive and light-weight, reduce start up costs and favour also portability. The simple prototype design of the interferometer presented in Figure 1.3 was devised under the additional condition of a limited number of the optical components in order to reduce operational complexity. It is based on the simple optical scheme of Figure 1.1 mounted on a rotating stage to allow displacement visualization along different directions. Major components are a 5mW-He/Ne laser, a microscope lens, two field lenses, three front surface mirrors and a rotating stage.

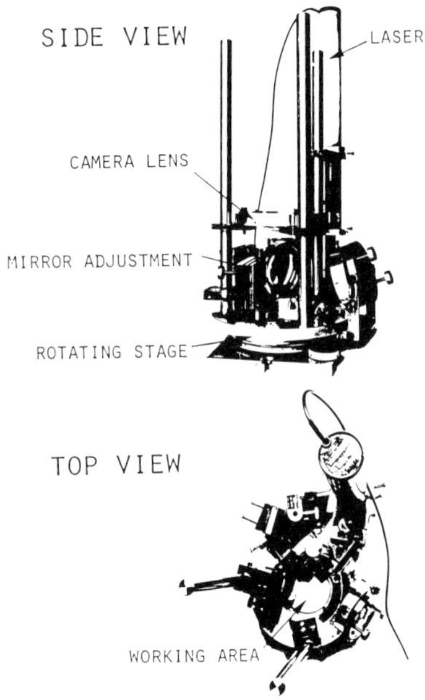

Figure 1.3

Residual stress computation

Moiré interferometry reveals directly, and with high sensitivity, in-plane displacement components. A theoretical residual stress vs. in-plane displacement formulation is, therefore, appropriate and is now outlined.

The superposition principle is invoked in order to solve the elasticity problem given by drilling a hole in plate originally subjected to biaxial residual stress field. The deformations to be measured are given by the introduction of a free surface in the unknown biaxial stress field. In order to obtain the stress field in the vicinity of the hole, two known ancillary stress solutions are superposed [2]. The first is given by the expression in polar coordinates of the unknown biaxial stress field, the second is the exact solution for a perforated infinite plate subjected to biaxial far field stresses [7]. Subtraction of the two solutions yields

$$\sigma_r = -\frac{a^2}{r^2}\frac{\sigma_1 + \sigma_2}{2} + \left(3\frac{a^4}{r^4} - 4\frac{a^2}{r^2}\right)\frac{\sigma_1 - \sigma_2}{2}\cos 2\Theta \tag{1.3}$$

$$\sigma_t = \frac{a^2}{r^2}\frac{\sigma_1 + \sigma_2}{2} - 3\frac{a^4}{r^4}\frac{\sigma_1 - \sigma_2}{2}\cos 2\Theta \tag{1.4}$$

in which $\underline{a}$ is the hole radius, σ_1 and σ_2 is the unknown biaxial residual stress field, r is the radial distance from the hole centre and Θ is the angular coordinate formed with the axis of traction σ_1 (positive in anti-clockwise direction). For a state of plane stress and for an isotropic and linear elastic behaviour of the material, the following final relationships between in-plane displacement components and the unknown residual stresses are obtained upon integration of the compatibility equations [8],

$$\upsilon_r = \frac{a^2}{E}\left\{\frac{1+\nu}{2r}(\sigma_1 + \sigma_2) + \frac{1}{2r}\left[4 - \frac{a^2}{r^2}(1+\nu)\right](\sigma_1 - \sigma_2)\cos 2\Theta\right\} \tag{1.5}$$

$$\upsilon_t = \frac{a^2}{E}\left[\frac{1}{r}(\nu - 1) - \frac{a^2}{2r^3}(1+\nu)\right](\sigma_1 - \sigma_2)\sin 2\Theta \tag{1.6}$$

Since Moiré interferometry conveniently yields the diametral change, S, at the hole rim, equation (1.5) becomes, upon introduction of the r/a = 1 condition,

$$S_\Theta = B(\sigma_1 + \sigma_2) - A(\sigma_1 - \sigma_2)\cos 2\Theta \tag{1.7}$$

where $A = (\nu - 3)a/E$ and $B = (1+\nu)a/E$. According to the generic coordinate system of Figure 1.4, the unknowns σ_1, σ_2 and α can be determined by evaluating the radial displacement along three directions (i.e. $\Theta_1 = -\alpha$, $\Theta_2 = -\alpha + 45°$ and $\Theta_3 = -\alpha + 90°$) and solving the system of three equations in three unknowns as proposed in [9]. The final equations are the following

$$\mathrm{tg}\ 2\alpha = \frac{S_1 + S_3 - 2S_2}{S_1 - S_3} \tag{1.8}$$

$$2\sigma_1 = \frac{S_1 + S_3}{2B} - \frac{1}{A \sin 2\alpha}\left[\frac{S_1 + S_3 - 2S_2}{2}\right] \quad (1.9)$$

$$2\sigma_2 = \frac{S_1 + S_3}{2B} + \frac{1}{A \sin 2\alpha}\left[\frac{S_1 + S_3 - 2S_2}{2}\right] \quad (1.10)$$

While the rather lengthy derivation of the previous equations is given elsewhere [8], equations (1.8 - 1.10) confirm those reported in [9].

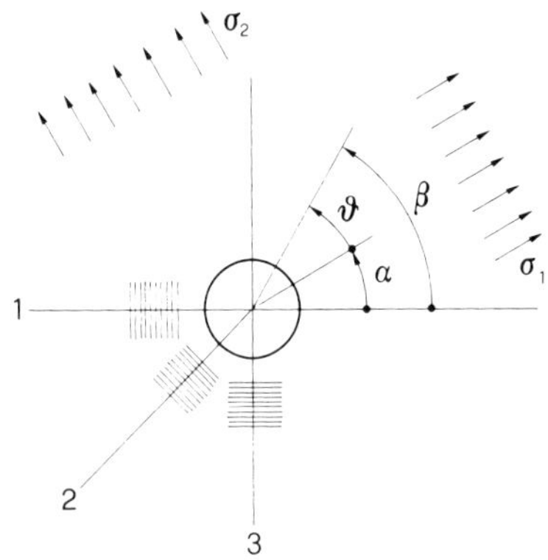

Figure 1.4

APPLICATION OF THE METHOD

In the most general case of residual stress system, three are the unknown variables (i.e. σ_1, σ_2 and α) to be determined. An initial verification of the viability of the method as far as specimen preparation and computation of stresses is presented here for a simpler stress system (i.e. $\sigma_1 = \sigma$, $\alpha \neq 0$, $\sigma_2 = 0$ in Figure 1.4).

After a preliminary annealing cycle to relieve forming residual stresses, reflective diffraction gratings with lines at 0°, 45° and 90° were replicated on the surface of mild steel bars (width W = 50mm, thickness t = 5mm, length L = 150mm). Reference lines marked in advance on the mould revealed that the angle between the direction of traction and the active direction of the grating 1 (i.e. α according to Figure 1.4) was equal to 28°. Holes (3mm in diameter) were carefully drilled through the gratings at the centre of the specimens. After hole drilling, each specimen was mounted on a loading rig and tensile far-field stresses up to 100 MPa applied while simultaneously visualizing the in-plane displacements by Moiré interferometry. Typical magnified fringe patterns relative to two directions are shown in Figure 1.5. They reveal remarkable fringe contrast and a successful drilling operation in mild steel.

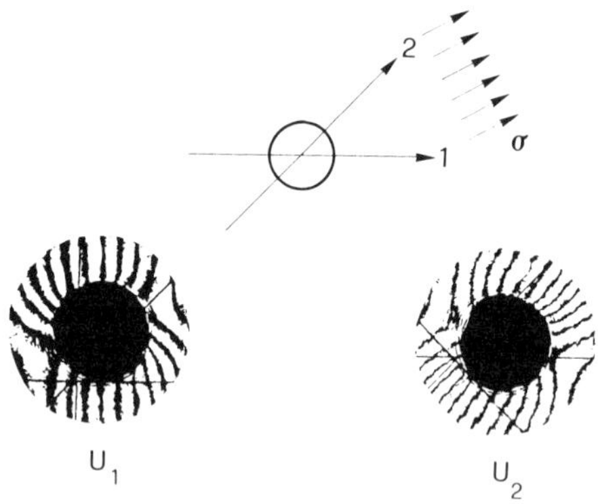

Figure 1.5

In order to obtain the magnitude of the applied stress and its direction with respect to the grating direction from the experimental evidence, only the stress field for a perforated plate in tension [7], was used in the derivation of the following equations yielding the radial and tangential displacements about the hole as a function of β as defined in Figure 1.4.

$$U_r = \frac{\sigma r}{2E}\left\{[(1-\nu) + (1+\nu)\frac{a^2}{r^2}] + [(1+\nu) + 4\frac{a^2}{r^2} - (1+\nu)\frac{a^4}{r^4}]\cos 2(\beta-\alpha)\right\} \quad (1.11)$$

$$U_t = -\frac{\sigma r}{2E}[(1+\nu) + 2(1-\nu)\frac{a^2}{r^2} + (1+\nu)\frac{a^4}{r^4}]\sin 2(\beta-\alpha) \quad (1.12)$$

Upon combination of equations (1.11) and (1.12), the general equation for the displacement component in the measurement direction can be written as follows

$$S = 2U_r \cos(\beta) - 2U_t \sin(\beta) \quad (1.13)$$

Two methods for the computation of the stress σ and the direction α were used based on two and a single fringe pattern, respectively.

(1) Two-picture method
This approach is equivalent to the one which resulted in equations (1.8 - 1.10), namely one picture for each unknown. Hence, two displacements across the hole for two directions, for example,

$U_r(\beta = 0°)$ and $U_r(\beta = 45°)$, hence only two fringe patterns, were required for the determination of the two unknowns σ and α. The results for two load levels of the resulting system of two equations in two unknowns are reported in Table 1.1. The correlation between the computed results and their actual values is satisfactory confirming the merit of the Moiré interferometric method.

Table 1.1

Actual and computed values of α *and* σ *as obtained by the Moiré interferometric hole-drilling method*

Two-picture method

Test	Actual		Computed	
	σ (MPa)	α (°)	σ (MPa)	α (°)
‡ 1	63.4	28	54	28.6
‡ 2	95.2	28	88	29.9

(2) Single-picture method
The previous approach to the determination of magnitude and direction of the applied stress, however, does not exploit the full-field nature of the Moiré interferometric information. An alternative method based on the determination of a single fringe pattern is the following one. For a/r = 1, equation (1.12) becomes

$$S = \frac{2a\sigma}{E} \left\{ [1 + 2 \cos 2(\beta-\alpha)]\cos \beta + 2 \sin 2(\beta-\alpha) \sin\beta \right\} \qquad (1.14)$$

The direction β_m associated with the maximum displacement component and its magnitude, S_m, in the 1-direction can be easily determined from inspection of the Moiré fringes as explained in Figure 1.6 with reference to an actual residual displacement pattern. Analytically, the unknown angle α is determined from the following condition

$$\frac{dS}{d\beta} = - 2 \sin 2(\beta_m-\alpha)\cos \beta_m - (1-2\cos 2(\beta_m-\alpha))\sin\beta_m) = 0 \qquad (1.15)$$

After α is obtained by solving equation 1.15 (i.e. its value is independent of the stress level), the magnitude of the applied stress, σ , is determined upon substitution in equation 1.14 of β_m and α and of the maximum relative displacement component, S_m. Inspection of the results obtained with this method which are reported in Table 1.2 demonstrates its remarkable accuracy. This one-picture approach is appealing since it simplifies the experimental portion of the method which is presently the most time-consuming and improves the utilization of the experimental evidence.

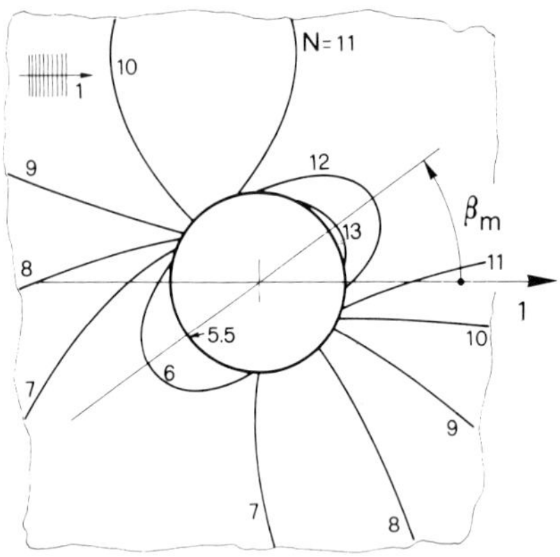

$$S_m = \frac{\Delta N}{f} = \frac{7.5}{2.4} = 3.13\ \mu m$$

Figure 1.6

Table 1.2

Actual and computed values of α and σ as obtained by the Moiré-interferometric hole-drilling method

Single-picture method

Direction 1 (see Figure 1.4)

Test	Actual		Experimental	
	σ (MPa)	α (°)	σ (MPa)	α (°)
‡ 3	31.7	28	29	26
‡ 4	63.4	28	68	25
‡ 5	95.2	28	103	29

Direction 2 (see Figure 1.4)

Test	Actual		Experimental	
	σ (MPa)	α (°)	σ (MPa)	α (°)
‡ 4	63.4	28	62	26
‡ 5	95.2	28	110	29

FINAL REMARKS

The Moiré interferometric hole-drilling method for residual stress measurements presented is characterized by full-field, high measuring sensitivity. The relatively simple specimen preparation procedure involving a diffraction grating replica and the design of a low-cost and portable instrument are expected to promote the use of the method for measurements in the field. Equations for the computation of stresses from in-plane displacement data have been presented.

The combination of the full-field nature of the method and the analytical stress vs. displacement formulation result in:

(1) full exploitation of the nominal sensitivity in the sub-micron range (0.417μm per fringe in the present application) since measurements are collected at hole rim where the maximum relaxation occurs;
(2) no preliminary gauge dependent calibration;
(3) relaxed accuracy in hole location since it does not affect stress computation;
(4) no a-priori definition of hole diameter, i.e. the optimal hole diameter can be determined by progressively enlarging an initially small hole;
(5) qualitative verification of the hypothesis of uniform residual stress distribution from analysis of the fringe patterns.

Although additional work is required before the method may be used in an industrial environment, the initial applications to a relatively simple stress system reported in this chapter have already demonstrated the potential of the method.

ACKNOWLEDGMENTS

The valuable contributions to the present study of T. Tedaldi and of the laboratory technicians are gratefully acknowledged. The research has been partially funded by Ministero della Pubblica Istruzione.

REFERENCES

[1] Almen, O., Black, P.H. (1963) Residual Stresses and Fatigue in Metals, McGraw-Hill Publ. Co., New York
[2] Soete, W., Vancrombrugge, R. (1950), "An Industrial Method for the Determination of Residual Stresses", Proc. SESA, 8(1), pp. 17-26
[3] Anon. (1981), "Standard Method for Determining Residual Stresses by the Hole-Drilling Strain-Gage Method", ASTM Standard E837-81, American Society for Testing and Materials, Philadelphia

[4] Post, D. (1986), "Moiré interferometry", Chapter 7, Handbook of Experimental Mechanics, A.S. Kobayashi Ed., Prentice-Hall Publ. Co., New York, in press

[5] Durelli, A.J., Parks, V.J., Moiré Analysis of Strain, Prentice Hall, 1970

[6] Flaman, M.T. and Herring, J.A., "SEM/ASTM Round-robin Residual Stress-Measurement Study - Phase 1", Experimental Techniques, Vol. 10, No. 5, 1986, 00. 23-25

[7] Timoshenko, S.P. and Goodier, J.N., Theory of Elasticity, 2nd ed., McGraw-Hill, New York, 1951, p. 80

[8] Nicoletto, G., "Development and Application of a New Semi-destructive Method for Residual Stress Measurement", Procs. National Conference of AIAS (Italian Soc. for Experimental Mechanics), 1987, in press (in Italian)

[9] McDonach, A., McKelvie, J., MacKenzie, P. and Walker, C.A., "Improved Moiré Interferometry and Application in Fracture Mechanics, Residual Stress and Damaged Composites", Experimental Techniques, Vol. 7, No. 6, 1983, pp. 20-24

2 Recent advances in neutron diffraction residual stress measurements in engineering components

H.J.MacGillivray,M.Bourke,G.A.Webster & P.J.Webster

SUMMARY

Recent improvements are described in the accuracy and resolution of the neutron diffraction residual stress measurement technique. Results are presented for near-surface residual stresses and compared with surface X-ray data.

INTRODUCTION

Increasingly in engineering applications, accurate measurements of residual stress are becoming of vital importance for predicting component lifetimes and failure probabilities. Several mechanical techniques are available including hole drilling and sectioning [1]. Most of these suffer from being either totally or at least partially destructive. Of the widely used methods only X-rays are truly non-destructive allowing measurements to be made on a sample which is then available for further experiments in an undamaged state.

The physical principles underlying X-ray and neutron radiation stress measurement are similar. The main difference arises from their relative penetrating powers and hence on the depth at which strain measurements can be made. Conventional X-ray techniques are used for establishing surface stresses; the radiation interacts electromagnetically with the electrons in the atomic shells and probes less than 100 microns into a material. Since this is the layer of surface grains, which under deformation are not compelled to maintain contact on all faces, measurements may not be truly representative of the bulk stresses. By comparison neutrons, because of their weak interaction with matter, can penetrate deeper and give stress measurements within the interior of a material typically several centimetres into steel or deeper for a low absorption material such as aluminium.

A number of research laboratories are currently developing neutron diffraction measurements of residual stress for a variety of applications in mechanical engineering. Work by the Imperial College/Salford University group has previously examined bulk stresses in the interior of samples such as welds and autofrettaged rings [2]. This chapter describes some preliminary measurements made on a plate of 7071 aluminium alloy shot peened on both surfaces, which is being used in a current British Society for Strain Measurement (BSSM) residual stress round-robin. A shot peened surface represents a severe test of the neutron diffraction method as it involves a steep stress gradient over a very short distance. This requires very accurate and well-defined collimation of the neutron beam to define a sampling volume with sufficient precision to allow the rapidly changing stresses to be measured. The measurements were made over two days on diffractometer D1a at the high flux reactor at the Institut Laue-Langevin in Grenoble. The efficacy of the technique is demonstrated and the problems encountered are discussed. Suggestions are made for further development of the measurement technique.

PRINCIPLES OF NEUTRON DIFFRACTION

In common with X-ray diffraction which has been employed since the 1950's, the neutron stress measuring technique detects the change in lattice spacing within the grains of a crystalline material resulting from a compressive or tensile applied stress. The measured strain may either result from a residual stress due to differential yielding in a loaded material, or from an applied load such as that encountered in a stress rig. As with other techniques, strains are measured which are subsequently converted to stresses. For both neutrons and X-rays the penetration is determined by the wavelength and elemental composition of the specimen material, with a typical neutron measurement involving a path length of up to 40mm compared with 50 microns for X-rays.

When a beam of monochromatic (uniform wavelength) neutrons falls on a crystalline material scattering occurs as defined by Bragg's law such that:

$$n\lambda = 2d \sin \theta \tag{2.1}$$

where n is the order of the diffracted peak
λ is the wavelength
d is the lattice spacing
2θ is the scattering angle in degrees

Differentiating whilst holding the wavelength constant gives:

$$2d\, \Delta(\theta) \cos \theta + 2\, \Delta d \sin\theta = 0 \tag{2.2}$$

then:

$$\varepsilon = \frac{\Delta d}{d} = -\Delta(\theta)\cot\theta \tag{2.3}$$

where ε is the strain of the lattice

If a crystal is subjected to a load then its unstressed lattice spacing d_0 will undergo a change $\Delta d = d - d_0$, giving the standard definition of strain in equation (2.3). This also implies a change $\Delta\theta$ in the neutron scattering direction (positive for compression, negative for tension) and is observed as a shift in the peak centre of the diffracted peak. Using a high resolution diffractometer it is possible to measure peak shifts as small as five thousandths of a degree which, depending on the cot θ term in equation (2.3), typically corresponds to 50 microstrain. This represents a stress resolution of about 4 MPa for aluminium and 10 MPa for steel. A more detailed description of the technique is given in Allen et al [3].

In order to specify the direction of strain measurement a scattering vector $\underline{Q}$ is defined by $\underline{Q} = \underline{k}_i - \underline{k}_f$ where $\underline{k}_i$ and $\underline{k}_f$ are the neutron scattering wave vectors describing the incoming and outgoing beams respectively. The strain is measured in a direction along vector $\underline{Q}$. This is illustrated in Figure 2.1 where the direction of strain measurement is normal to the peened surface. Note that in equations (2.1) - (2.3) θ is actually half the scattering angle Ø.

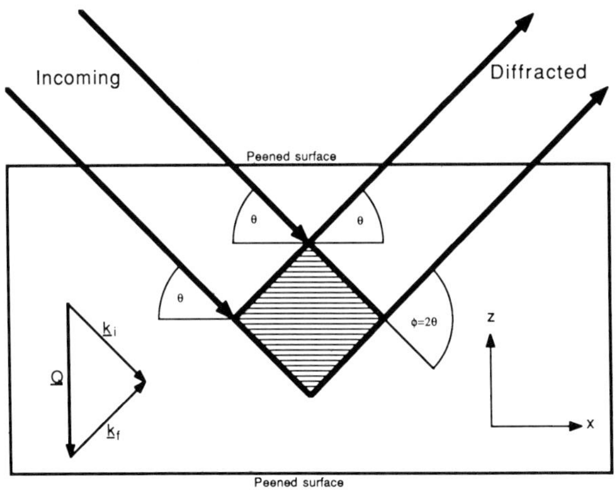

Figure 2.1: Definition of Q Vector

A metal is a polycrystalline agglomerate of many grains which, in the absence of texture, are randomly orientated with respect to any particular axis. Diffraction techniques sample those grains which by chance are aligned to give a scattered beam of neutrons in a direction dictated by Bragg's law, thereby averaging the strain from many grains in the sampling volume. When performing a measurement the incoming beam which is originally 20mm square is masked using a cadmium aperture. Cadmium is used because of its extremely high neutron absorption cross section.

Similarly, thin Cd slits are used to define the outgoing beam. As scattered radiation emerges in a cone no gain is made by masking in two directions on the outgoing beam. This procedure identifies a sampling volume fixed in space and corresponding to the intersection of the beams defined by the incoming and exit masks. The detector can only receive neutrons which are scattered from this volume. In most experiments the most crucial factor is an accurate definition of the sampling volume. Different specimen geometries require differently shaped sampling volumes. For example mapping the strain distribution around a crack tip requires an input beam 2mm square combined with an exit slit 2mm wide to give a sampling volume of $8mm^3$ (a more complete description of the different volume geomeries may be found in [4]). In situations where a degree of symmetry exists the shape of the sampling volume may be adjusted to take advantage of the symmetrical stress distribution.

Having defined the sampling volume in space, the position of the Bragg peak is measured by moving the neutron detector in small angular steps, scanning typically to one degree on either side of the peak centre, then a Gaussian is fitted to the recorded data to obtain the peak centre, height and width. The shift in peak centre from the nominal unstressed peak centre position is used to determine the average lattice strain for those grains in the sampling volume oriented in a particular direction. Generally the single crystal strain response is anisotropic for different directions in the lattice hence it is important to choose a Bragg reflection whose behaviour accurately reflects the macroscopic response of the material.

Strain measurements are made at different positions in the specimen by moving it in incremental steps through the sampling volume and the direction of strain measurement is altered by varying the specimen orientation with respect to the beam. When the principal stress directions are known only three orthogonal measurements are needed at each point to completely define the stress. On the high resolution diffractometer D1a at the Institut Laue Langevin, sample movement is performed using three stepper motor driven linear stages which may be mounted together or individually. This allows movement in three orthogonal directions (Figure 2.2). By using the diffractometer omega table it is possible to add a rotation about a vertical axis. In principle the positioning accuracy of the motors is better than 0.1mm but practical difficulties of sample location with the surveying equipment used at the moment reduce the absolute positional accuracy to about ±0.2mm.

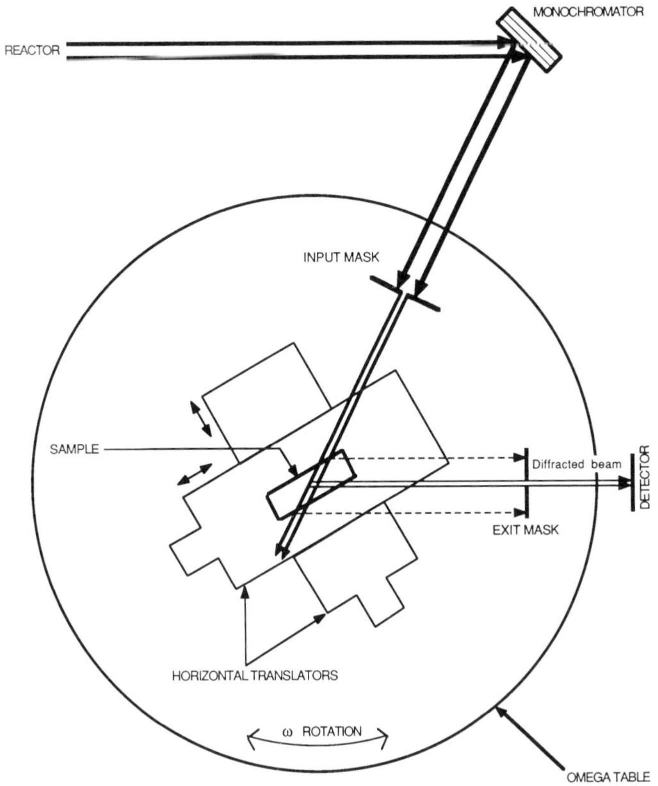

Figure 2.2: Experimental Set-up, D1a

SHOT PEENED SAMPLE

Shot peening is used to improve fatigue strength and inhibit surface crack initiation, stress corrosion cracking and intergranular corrosion. It is a cold working technique in which the surface is yielded by bombardment with closely sized spherical shot to produce compressive residual stresses in the surface. The specimen examined was a plate 75mm square x 10.6mm thick of aluminium alloy 7071. The complete specification of the alloy and treatment are given in appendix 2.1. It forms part of the BSSM Residual Stress Technical Group round-robin test programme for comparing measuring techniques. Neutron measurements were made in only two directions corresponding to an in-plane X measurement and a normal Z strain measurement (Figure 2.1). The Bragg reflection employed was the (311) which, with neutrons of wavelength 0.1905nm, gives a scattering angle of approximately 102.8 degrees. Each strain direction was measured at a series of through-thickness positions with particular emphasis being given to the first 1mm from the surface where measurements were made in steps of 0.1mm while in the interior 1mm steps were used.

Because the shot peening process results in a uniform strain profile in the plane of the specimen it was possible to use an elongated sample volume to increase the neutron scattering volume and improve the count rate, peak fitting statistics and accuracy. Maximising the volume of sample in the beam without compromising resolution is particularly important when working with weakly scattering materials like aluminium. Figure 2.3 shows how a 30mm x 1.5mm input mask was combined with a 2mm exit mask to define a sampling volume of 30mm x 1.5 x 2mm through which the sample was moved with the long dimension parallel to the surface.

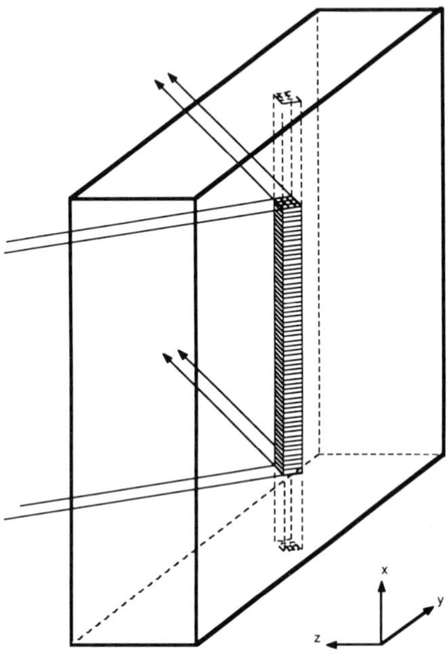

Figure 2.3: Sampling volume geometry

Shot peening produces a steep surface stress gradient and large changes in strain were observed as the sampling volume progressively penetrated the material. Since the changes were occurring over a distance comparable to or smaller than the depth of the sampling volume it was particularly important to deconvolute the position data to correct for the partial immersion of the sampling volume. The need for deconvolution is illustrated in Figure 2.4 which shows the progressive immersion of the sampling volume as it enters the surface of the peened plate. The dimension AD (Figure 2.4a) is approximately 2mm which means that the peening effect will have substantially reduced before complete immersion has taken place. In Figure 2.4b the sampling volume is partially buried in the surface and neutrons are diffracted to the detector from the triangle FHI. However this effectively samples more material close to the surface than close to point 1. Consequently the position of measurement is taken to occur at the centre of gravity of the shaded region on a line GI at a point indicated by the arrow. Similarly in Figure 2.4c where complete immersion has taken place sampling is taken to occur at the intersection of JM and KL. Due to this progressive immersion it was found that, before deconvolution, intensity was recorded at a distance of 0.5AD on either side of the physical limits of the sample.

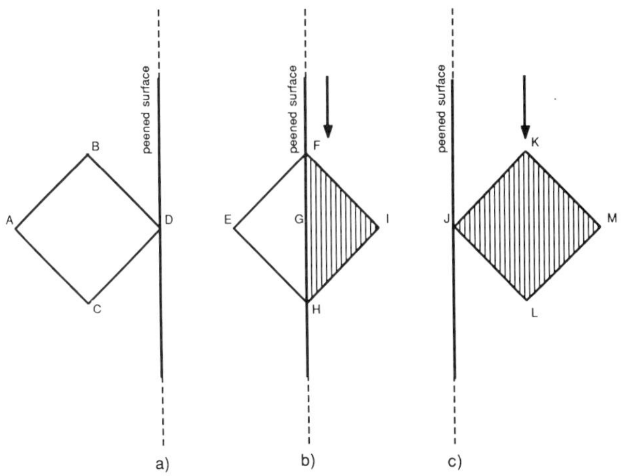

Figure 2.4: Progressive immersion of sampling volume

RESULTS

The deconvoluted peak centres and heights are given in Figure 2.5. It should be noted that the scan did not extend completely through the sample to emerge from the back face B, since the recorded peak intensities do not fall to zero at this face. The position of the peak centre is directly related to the average strain where a peak shift to a larger angle corresponds to a reduction in the value of the lattice spacing (equation 2.2) and thus to a compressive strain at that position. Conversely a shift to a smaller angle indicates a tensile strain in the material.

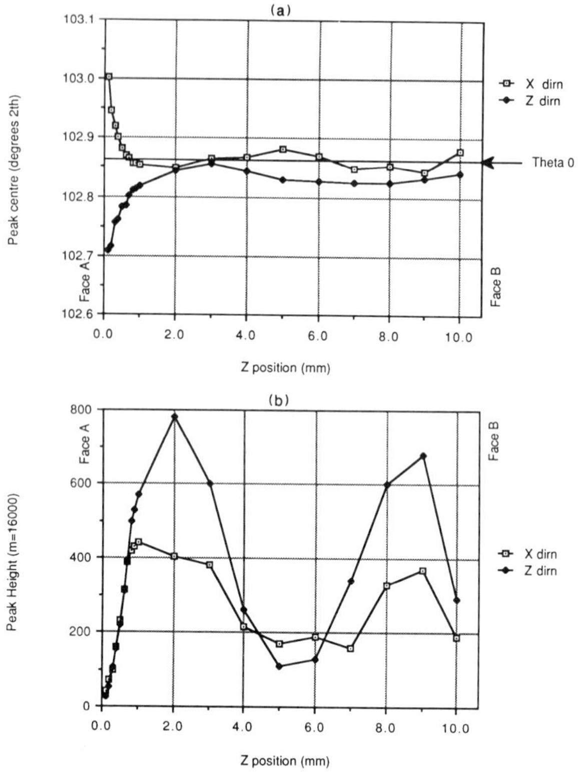

Figure 2.5: (a) *Measured Peak Shifts vs. Depth into plate*
(b) *Peak Heights vs. Depth into plate*

In order to determine the absolute value for a peak position θ_0 corresponding to d_0 for material in a relaxed unstressed state, it is necessary to measure the scattering angle for a sample of stress relieved material. However when this is not available it is possible to invoke equilibrium such that the planar stress must average zero for a complete section through the Z direction. By invoking equilibrium and assuming a mirror image for strains on the front and back surfaces, a value for θ_0 may be determined. For the stress calculation the assumption was made thatthe strains in the X direction were of equal magnitude and sign to the Y direction. The standard elastic stress-strain relationships given in Appendix 2.2 were then applied. The calculated stress values are shown in Figure 2.6.

X-ray measurements were made using Phillips equipment at Rolls Royce Derby. Surface stresses were measured in the X and

Y directions on both peened faces of the specimen using two radiations Cu Kα and Cr Kα. By using copper and chromium radiation of differing wavelength it is possible to vary the X-ray penetration depth. The chromium is less penetrating and approximately 95% of the radiation is diffracted within 27 microns of the surface. For copper radiation approximately 95% of the diffraction takes place within 84 microns of the surface. The X-ray measurements are included in Appendix 2.3 and an average of the X and Y stresses is plotted for side A (Figure 2.6b) at a position corresponding to half the 95% diffraction limit.

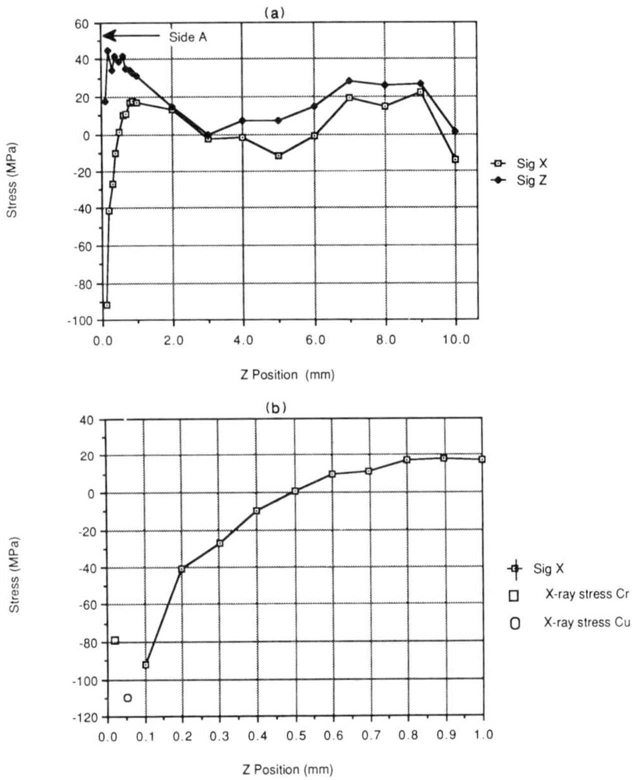

Figure 2.6: (a) *Stresses planar and perpendicular*
(b) *Planar Stress 0-1mm*

DISCUSSION AND CONCLUSIONS

In the analysis it has been assumed that the stress in the X and Y directions are equal and θ_0 has been chosen to satisfy equilibrium. When these conditions are satisfied Figure 2.6 indicates that the stress perpendicular to the free surface σ_Z at side A is approximately zero as is required at any surface. There is the same tendency at side B although this side was not examined in such detail. Observations of zero stress perpendicular to a free surface can therefore be used as an independent check of the technique.

Figure 2.6 shows that compressive residual stresses developed by the shot peening process are limited to a depth of about 0.5mm. From Figure 2.4 it can be seen that the shallowest stress measurement corresponds to a depth below the surface of 0.1mm and a compressive stress of 96MPa. In comparison the yield stress of the material is approximately 460MPa giving a maximum residual stress of about 20% of yield.

The X-ray results vindicate the assumption that the X and Y strains in the plane of the plate are equal. Also the recorded stresses with the more penetrating chromium radiation are in good agreement with the neutron measurements. In particular a comparison of the stresses produced by the more shallowly penetrating copper radiation with those from the chromium radiation suggest the existence of a plateau in the stress distribution within about 100 microns of the surface. This may imply an approximate plane stress condition as indicated by the low value of σ_2 (Figure 2.6a) in this region. When comparing the absolute stress values and their positional location it is important to note the vulnerability of both the neutron and the X-ray techniques to selection of an appropriate bulk elastic modulus. No allowance was made for the inherent variation in penetration depth due to the variation in angle of incidence during a single X-ray stress measurement.

These initial measurements are not comprehensive and further refinement of the neutron technique is possible to improve resolution near the surface. In particular by looking at the strain at 45 degrees to the surface it is hoped that some overlap may be obtained between the X-ray and neutron techniques. Also there is evidence of preferred orientation in the plane of the plate due to the marked neutron intensity variation seen through the thickness of the plate (Figure 2.5) when the sampling volume was completely buried. Micro hardness tests and polishing/etching processes would be useful in understanding the nature of the texture variation through the thickness of the material. Nevertheless it is believed that it has been demonstrated that the neutron diffraction technique can be employed to measure residual stresses in Aluminium alloys to an accuracy of ± 10MPa in regions of steeply changing stress.

ACKNOWLEDGMENTS

The authors would like to thank Dr. A. Hewat and Prof. K.R.A. Ziebeck for their help at the ILL. Thanks are also due to Mr. J. Davies the D1a technician whose assistance was invaluable. The data processing at the ILL was performed by K. Low and G. Mills of Salford University. Finally we would like to thank Mr. C. Baxter and Dr. P. Webster of Rolls Royce Derby for making the X-ray measurements.

REFERENCES

[1] Beaney, R.M. and Proctor, E. "A Critical Evaluation of the Centre Hole Drilling Technique for the Measurement of Residual Stresses". Strain, Vol. 10, 1974, pp. 7-14

[2] Stacey, A., MacGillivray, H.J., Webster, G.A., Webster, P.J. and Ziebeck, K.R.A., "Measurement of residual stresses by neutron diffraction". J of Strain Analysis, Vol. 20, No. 2, 1985, pp. 93-100

[3] Allen, A.J., Hutchings, M.T. and Windsor, C.G., "Neutron diffraction methods for the study of residual stress fields", Advances in Physics, Vol. 34, No. 4, 1985, pp. 445-473

[4] Smith, D.J., Webster, G.A. and Webster, P.J., "Measurement of residual stress and the effects of prior deformation using the neutron diffraction technique", Welding Institute Research Report, 338/1987.

Appendix 2.1

Shot peened 7071 aluminium alloy plate 75mm x 75mm x 10.6mm

Nominal Composition	4.5% Zinc
	2.5% Magnesium
Heat Treatment	T651 aged beyond peak strength
Young's Modulus (E)	69GPa
0.2% Proof Stress	460MPa
Elongation at Fracture	12%
Poisson's Ratio (ν)	0.3
Peening Conditions	S 110 shot
Intensity	.008" - .012"
Coverage	125%

Appendix 2.2: *Neutron Diffraction Stress Analysis*

$$\sigma_x = \frac{E}{(1+\nu)(1-2\nu)}[(1-\nu)\varepsilon_x + \nu(\varepsilon_y+\varepsilon_z)]$$

$$\sigma_y = \frac{E}{(1+\nu)(1-2\nu)}[(1-\nu)\varepsilon_x + \nu(\varepsilon_y+\varepsilon_z)]$$

$$\sigma_z = \frac{E}{(1+\nu)(1-2\nu)}[(1-\nu)\varepsilon_z + \nu(\varepsilon_x+\varepsilon_y)]$$

where ν = Poissons ratio
E = Young's modulus
x, y, z are orthogonal strain directions indicated in Figure 2.3

Appendix 2.3: *Rolls Royce X-ray measurements on A17071 Shot Peened sample*

	Side A	Side B
Chromium Kα Radiation:		
X direction	-78	-94
Y direction	-81	-98
Copper Kα Radiation:		
X direction	-107	-126
Y direction	-112	-128

units of stress MPa

3 Moiré study of 2 and 3-hole arrays of composite material fasteners

P.J.Herrera-Franco, G.L.Cloud & K.Zimmerman

SUMMARY

Strain distributions in the vicinity of multiple-pin connections in FGRP composite were measured using three-axis high-sensitivity Moiré interferometry and digital data reduction. Surface strains were measured in three different directions so that the rosette equations can be used to obtain maximum strains. Two-hole and three-hole patterns were examined for ideal and worst-case load distributions. Stress concentration factors near the holes were obtained and interactions between holes were studied. Of the six cases examined, lowest stresses occur for the 2-hole parallel pattern, while highest stresses are found for the nonideal 2-hole tandem pattern where most of the load is carried by one hole because of misalignment that can easily occur at assembly.

OBJECTIVES AND SCOPE

In this study, high-sensitivity three-axis Moiré interferometry was used to obtain the stress and strain distributions in the vicinity of hole arrays. The purposes of these experiments were: (1) to determine the stress and strain distributions around the loaded holes; (2) to study the interactions between the holes; (3) to determine the effect of the unloaded hole upon the stress and strain distribution when only one hole is loaded.

One of the configurations considered had two holes in a row parallel to the load direction. An ideal solution to this statically indeterminate problem, where a goal is to optimize joint performance, is for each pin to carry one half of the applied load. This ideal situation was simulated using cables and pulleys to divide the load. However, in less ideal conditions, some hole misalignment can exist, thus producing a nonuniform distribution of load between the two pins. In order to simulate the worst condition, the holes were loaded individually in turn.

In the case of two holes in a row perpendicular to the direction of the load, both holes were loaded as equally as possible by striving for symmetry in the specimen and loading system.

For the three-hole pattern, three cases should be considered, namely: (1) all three holes loaded equally; (2) the lower two holes loaded equally, and the upper single hole free of load; (3) the single upper hole loaded and the two lower holes left unloaded. In this study, only the equiloaded case was examined for the three-hole staggered array.

The material tested was glass-reinforced epoxy (R1500/1581, 13 plies, 3.5mm thick) supplied by Ciba-Geigy, Fountain Valley, CA 92708, USA. The "fasteners" were pieces of 6.25mm (.25in.) dia. drill rod which were a snug push fit in their holes.

BACKGROUND

The prediction of stress and strain in the vicinity of pin-loaded holes is necessary in order to better understand the behaviour of composite joints and in order to utilize the full potential of these materials as structural elements.

Several investigators have developed analytical procedures for calculating the strength of bolted and pinned joints in composites. The majority of these studies have centred on the stresses around a single hole loaded by a rivet, pin or bolt. There exist also a few papers on studies of multiple hole arrays. Chang and Scott [1] presented a method for predicting the failure strength and the failure mode of fibre-reinforced composite laminates containing one or two pin-loaded holes. They also presented [2] a method for sizing composite laminates containing one, two or several pin-loaded holes, the holes being arranged either in single row or in two parallel rows. Wong and Matthews [3] presented a finite element analysis of bolted CFRP. Pyner and Matthews [4] studied the strength of single and multi-hole joints in GFRP plates bolted to a steel fitting. They suggested that the joint strength expressed as load/bolt decreases as the joint geometry becomes increasingly complex. Hyer and Liu [5] used photoelasticity to study stresses in the inner laps of double lap isotropic joints. Oplinger [6] studied analytically and experimentally parallel and series fastener arrays for several types of materials. Cloud and several coworkers [7] report analytical and experimental results for 1-, 2-, and 3-pin joints in FGRP and they also present a comprehensive survey of literature on mechanical fastening of composites.

EXPERIMENTAL PROCEDURE

The Moiré technique used is based on the extensive developments described by McKelvie and Walker et al [8, 9] and by Post et al [10, 11, 12]. It has been further developed and analysed by Herrerra [7, 13]. In its current form, the interferometer yields displacement and strain data for 3 orientations (0°, 45°, 90°) over a field of up to 15cm dia. Micrometer adjustments of the mirrors allow sensitivity and pitch mismatch to be adjusted over a broad ranges.

The specimens used in the experiments and the nomenclature are illustrated in Figure 3.1. The spacing between the centres of the holes was chosen so that d/D = 2.5 and some interaction between the holes could be expected [2]. To ensure accurate alignment of the holes and identical hole placement and hole diameters in each lap, all three laps of the specimens were machined simultaneously as a sandwich. The long sides of the specimens were machined parallel to each other, and the two rows of small holes were oriented parallel and perpendicular to the long edges of the specimen.

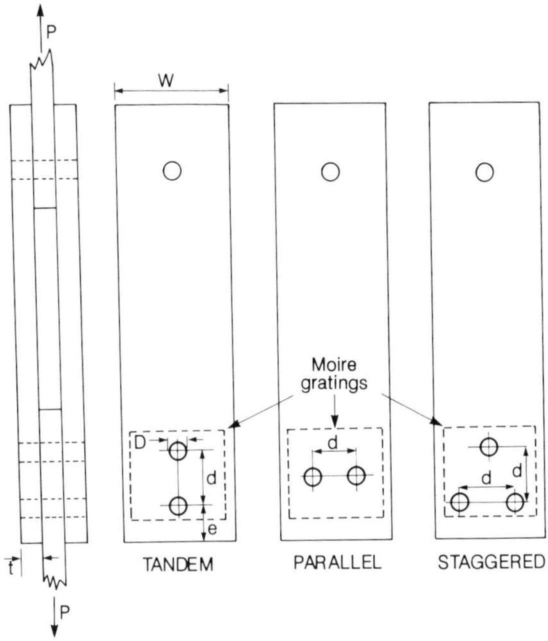

Figure 3.1: Joint Geometry and Nomenclature

High-frequency reflecting Moiré gratings were formed on the surface of each specimen [13, 14]. When the specimens are placed in the interferometer, Moiré fringe patterns of displacements are obtained by overlapping beams of light diffracted from the grating as it is illuminated by coherent beams of laser light. The patterns are checked using a PC-based digital image processing system in real-time mode. They are recorded either on floppy disc or photographically using Kodak Technical Pan film. The fringe locations for various chosen axes are then digitized. Strain contours were obtained using a Moiré data reduction computer program developed by Cloud and coworkers [15]. The approach uses piece-wise spline fitting with least-squares plus numerical

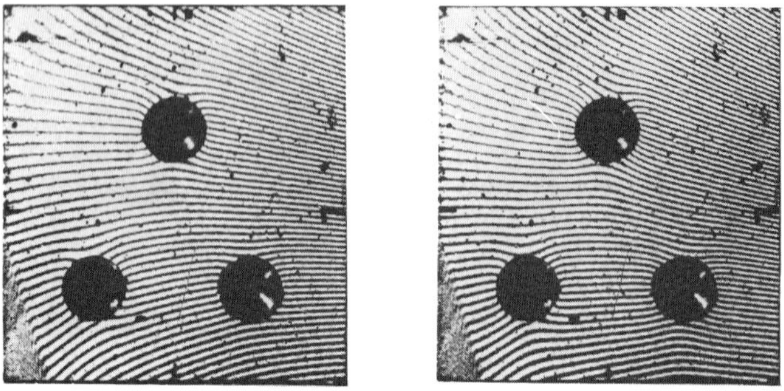

Figure 3.2: *Moiré Fringes of Displacement Parallel to Direction of Load for Specimen with Three-Hole Staggered Array*

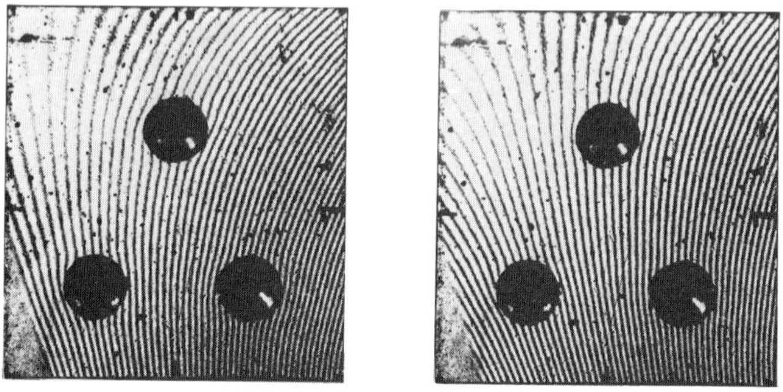

Figure 3.3: *Moiré Fringes of Displacement Perpendicular to Direction of Load for Specimen with Three-Hole Staggered Array*

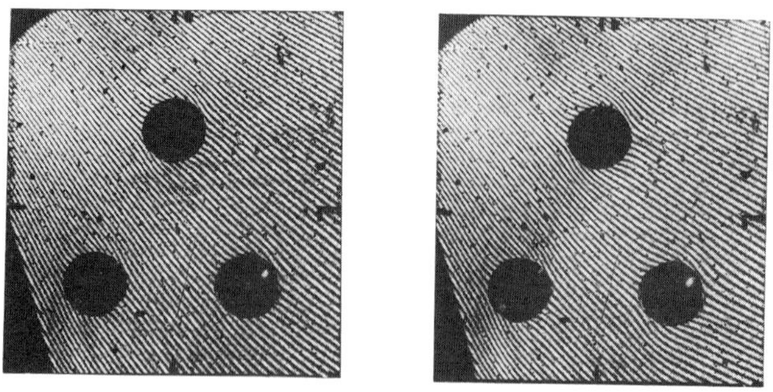

Figure 3.4: *Moiré Fringes of Displacement 45-deg. to Direction of Load for Specimen with Three-Hole Staggered Array*

differentiation to obtain strain. Data are always handled in pairs, such as between "no load" and "at-load", so that the processing is in the preferred differential mode.

TYPICAL MOIRÉ RESULTS AND ANALYSIS

Each of the six cases studied yielded a minimum of six Moiré patterns which in turn yielded about 40 plots which show strain for several axes in the three directions for the whole field examined. Space limitations preclude presentation of even a significant fraction of the Moiré data, which contain much more information than has been used in this study, or of the strain results, which lead to specific observations about the mechanics of fasteners in the configuration examined.

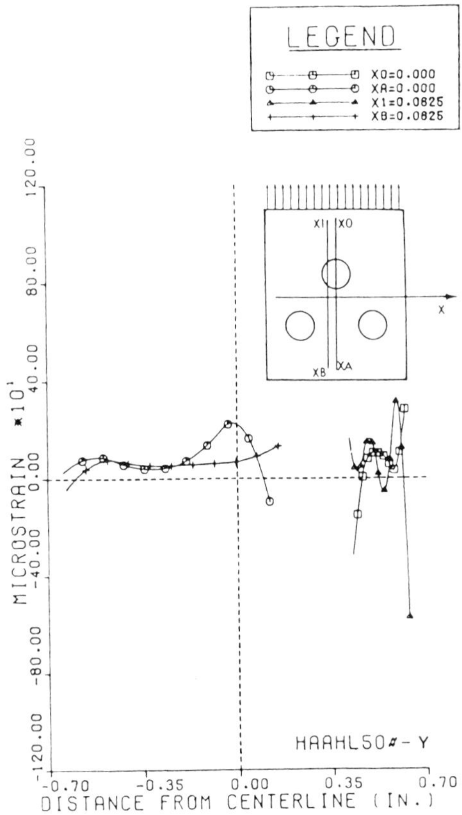

Figure 3.5: Strain ε_y along Lines Parallel to Direction of Load

Presented below are some typical results for only one strain component for only one of the cases, that of the three-hole staggered array, along with some interpretive remarks. Representative results and implications for other cases can be found in related reports and papers [7, 13, 16, 17, 18, 19].

For the three-hole configuration, all holes were loaded equally via a cable and pulley arrangement. Figure 3.2 shows a typical Moiré fringe pattern for displacement parallel to the direction of loading. Figure 3.3 shows fringes corresponding to displacements transverse to the load axis, and Figure 3.4 shows 45° displacements.

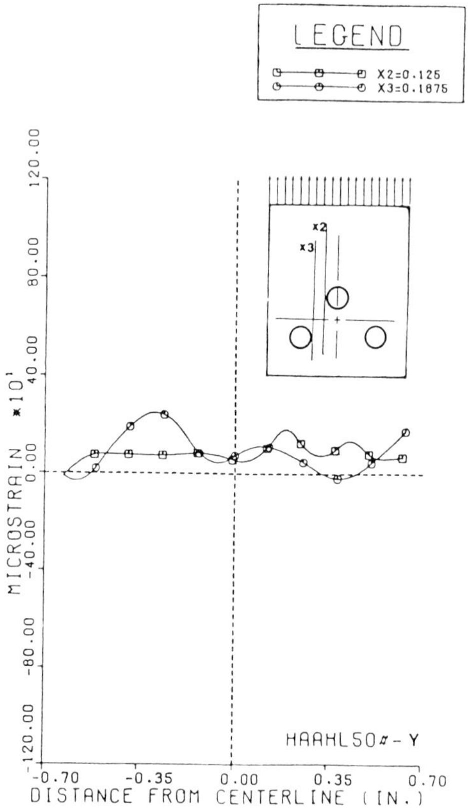

Figure 3.6: Strain ε_y along Lines Parallel to Direction of Load

Figure 3.5 shows strain ε_y for lines located above and below the upper hole. As expected, the strain above the hole is tensile, except close to the upper edge, as has been observed for the single-pin case. For the two axes in the bearing region of the upper hole a very interesting behaviour is observed; ε_y is negative only very close to the edge of the upper hole, and then it changes to a peak tensile strain which then decreases to an average value. Figures 3.6 and 3.7 show strain ε_y for axes which are tangent to the upper hole and also to the lower ones. In Figure 3.6, line x2 = 0.125 is tangent to the upper hole and x3 = 0.1875 is tangent to the inner edge of the lower left hole.

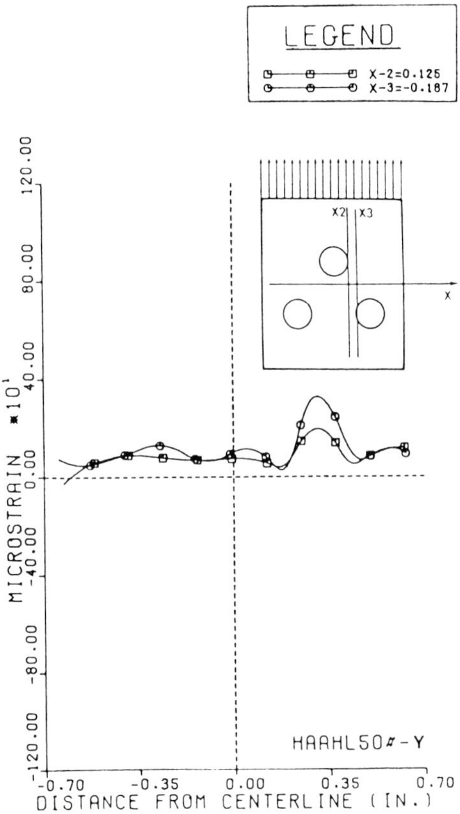

Figure 3.7: Strain ε_y along Lines Parallel to Direction of Load

In the plot for x2 there are two peak values, at y = 0.1818 and at y = 0.410. These two peaks correspond to positions which are above and below the horizontal diameter of the upper hole. Then, ε_y becomes uniform close to the lower holes. In axis x3 = 0.1875, the peak occurs at y = -0.307, which is slightly below the horizontal diameter of the lower left hole. For both axes, there is a peak strain at the horizontal diameter of the upper hole and a peak also occurs near the horizontal diameter of the lower right hole.

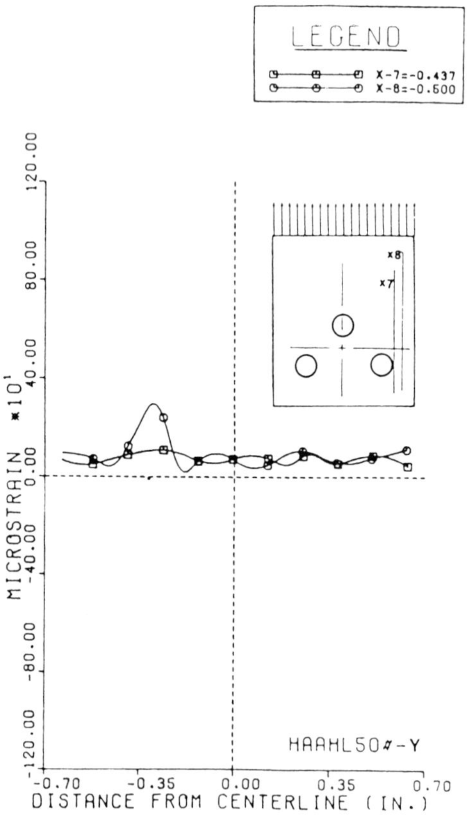

Figure 3.8: Strain ε_y along Lines Parallel to Direction of Load

Figure 3.8 shows the ε_y in the ligament region between the lower hole and the specimen long edge. Again, the location and value of the peak strain near but not exactly coincident with the horizontal diameter of the lower right hole are noteworthy.

There was a lack of symmetry, amounting to about 8%, between the peak strains at the inner edges of the two lower holes. For the upper hole, the difference between the two edges is very small.

The strain distributions show only a small bearing region for the upper hole, and below it the strain is positive. High strain gradients due to interaction of the three holes are not noticed. Further study is needed to elucidate the effect of hole spacing on strain distribution. Also, in the region between the two holes, the ε_y is tensile, and it should tend to cancel the compressive bearing strain from the upper hole.

DISCUSSION

Peak stress concentration factors (SCF) were obtained from the strain plots for important locations in each of the configurations studied.

The SCF at the edge of each hole was computed by dividing the stress in the direction of the load by the far-field stress defined as σ_0 = P/wt. The SCF was calculated at points located at the edge of each hole on the horizontal diameter for the 2-hole tandem array. It was calculated at both edges of each hole on the horizontal diameter for the case of two holes in parallel. For the 3-hole array, the SCF was obtained at both edges on the horizontal diameter of the upper hole and also at the inner edges of the two lower holes. These results are compared with the results for the single-pin case [13]. All these results are summarized in Tables 3.1 - 3.3.

Table 3.1

Comparison of stress concentration factors

	TWO HOLE TANDEM			SINGLE HOLE
	both loaded	up loaded	lo loaded	
upper	5.57	12.1	4.97	14.0
lower	1.91	1.0	9.16	14.0

Table 3.2

Comparison of stress concentration factors

TWO HOLE PARALLEL		SINGLE HOLE
left hole	right hole	
0.65	1.16	14.0

Table 3.3

Comparison of stress concentration factors

THREE-HOLE ARRAY				SINGLE HOLE
upper hole edge		lower-left hole	lower-right hole	
left	right			
2.17	2.32	2.95	1.54	14.0

When comparing the results for the different cases, keep in mind that the normalization is on the basis of the gross far-field stress, and the division of gross load among separate pins would need to be accounted for separately if it is important and if it could be ascertained. This factor, as well as some others, is apparent in the SCF for the 2-hole parallel array, where a value less than one is noted. The average SCF for this case would be 1.8 if it is assumed that the load is divided equally between the holes and the SCF calculated on the basis of the "far field stress per hole". Such a basis is not adequate for overall comparisons of joint performance, although it might be useful in visualizing the detailed mechanics of joint behaviour.

In terms of joint performance, then, the advantages of using multiple fastener arrays are evidenced by the reductions of stress concentration at each critical location in comparison with the SCF of 14 for the single pin. Even on the basis of load per fastener, there are advantageous gains in SCF, probably owing to the general softening of the joint area by added holes as well as to some fortunate interactions between holes which decrease some stress components in some areas. The worry here is that there might be unfortunate combinations at locations other than those for which the SCF was obtained. Detailed study of the entire strain field seems appropriate.

The parallel arrangement is seen to be the best in terms of reduced stress. The tandem configuration is worst, next to the single-pin joint, even in the ideal case of equal load division.

For the 2-hole parallel array, there is an increased reduction of the net section as compared with the tandem array, and larger SCF was expected. In fact, the stress distribution on the critical section for the parallel array was more uniform than expected, and decreased peak values along with a reduced SCF are found. This behaviour is probably a result of the reduction of joint stiffness along with a fortunate choice of hole spacing and edge distance.

Despite symmetry of loading, considerable lack of symmetry is apparent in the strain plots and in the SCF's for the 2-hole parallel array. A similar comment may be made about the results for the lower 2 holes (parallel pair) of the 3-hole array. These cases should be studied again with an improved loading arrangement. In the meantime, the average values for the two parallel holes in each case where the asymmetry is found is probably accurate for design purposes.

Some other interesting aspects of the stress-strain distributions around the pin-loaded holes in multiple-hole arrays can be visualized from careful examination of the many plots and Moire patterns produced. For example, in the case of the 2-hole tandem array when both holes are loaded equally, the resulting stresses at the net sections are not in the ratio of 2:1. In the case of the tandem array and only one hole loaded at a time, the resulting stresses are the result of interaction of the loaded hole and the unloaded hole. When the lower hole is loaded, the stress field in the region between holes is the sum of two positive stress fields. When the upper hole is loaded, its bearing stress tends to counteract the tensile stress field above the lower unloaded hole in the region where the effects overlap. A related view is that the lower hole softens the area below the loaded upper hole and thereby reduces bearing stress while perhaps increasing ligament stress.

These results suggest that stress reduction by drilling extra unloaded holes might be especially effective with composites.

Another aspect of the result which should be mentioned again is that the maximum stress did not occur at the ends of the horizontal diameter of the hole. Frocht and Hill [20] and others showed that this is always the case for snug-fitting pins in single-pin connections in isotropic materials. Evidently it is also true for the orthotropic material studied here.

REFERENCES

[1] Chang, F.K. and Scott, Richard,"Failure of Composite Laminates Containing Pin Loaded Holes-Method of Solution", J. Composite Materials, Vol. 18, May 1984, pp. 255-278

[2] Chang, F.K. and Scott, Richard, "Design of Composite Laminates Containing Pin Loaded Holes", J. Composite Materials, Vol. 18, May 1984, pp. 279-289

[3] Wong, C.M.S. and Matthews, F.L., "A Finite Element Analysis of Single and Two-hole Bolted Joints in Fibre Reinforced Plastic", J. Composite Materials, Vol. 15, September 1981, pp. 481-491

[4] Pyner, G.R. and Matthews, F.L., "Comparison of Single and Multi-Hole Bolted Joints in Glass Fibre Reinforced Plastic", J. Composite Materials, Vol. 13, July 1979, pp. 232-239

[5] Hyer, M.W. and Liu, D.H., "Photoelastic Determination of Stresses in Multiple-pin Connectors", Experimental Mechanics, Vo. 23, No. 3, September 1983, pp. 249-256

[6] Oplinger, D.W., "On the Structural Behaviour of Mechanically Fastened Joints in Composite Structures", Fibrous Composites in Structural Design, Edited by E.M. Lenoe, D.W. Oplinger and J.J. Burke, Plenum Press, 1978

[7] Cloud, G., Sikarskie, D., Vable, M., Herrera Franco, P. and Bayer, M., "Experimental and Analytical Investigation of Mechanically Fastened Composites", Technical Report No. 12844, U.S. Army Tank-Automotive Command, TACOM, Warren MI 48090, 400pp. Jan. 1987

[8] Walker, C.A. and McKelvie, J. "A Practical Multiplied Moiré System", Experimental Mechanics, Vo. 8 (8), 316-320, (August 1978)

[9] McDonach, A., McKelvie, J., and Walker, C.A., "Stress Analysis of Fibrous Composites using Moiré Interferometry", Optics and Lasers in Engineering, V. 1, 1980, pp. 85-105

[10] Post, D., "Optical Interference for Deformation Measurements", Mechanics of Nondestructive Testing, Plenum Publishing Corp., New York, 1980

[11] Czarnek, R. and Post, D., "Moiré Interferometry with 45 Deg. Gratings", Experimental Mechanics, Vol. 24 (1), March, 1984, pp. 68-74

[12] Post, D. and Baracat, W.A., "High Sensitivity Moiré Interferometry - A Simplified Approach", Exp. Mech., Vo. 21, No. 3, pp. 100-104 (Mar. 1980)

[13] Herrera-Franco, P.J., "A Study of Mechanically Fastened Composite Using High Sensitivity Interferometric Moiré Technique", Ph.D. Thesis, Michigan State University, Department of Metallurgy, Mechanics and Materials Science, 1985

[14] Basehore, M.L., and Post, D.,"High Frequency, High Reflectance Transferable Moiré Gratings", Experimental Techniques, Vol. 8, No. 5, May 1984, pp. 29-31

[15] Cloud, G.L., "Residual Surface Strain Distributions Near Holes Which are Cold Worked to Various Degrees", Technical Report AFML-TR-78-153, Wright-Patterson AFB, Ohio (1980)

[16] Herrera-Franco, P.J. and Cloud, G.L., "Strain-Relief Inserts for Composites Fasteners", Advanced Composite Conference, American Society for Metals, Dearborn, Michigan, December 1985

[17] Herrera-Franco, P.J., Cloud, G.L., "Experimental Analysis of Multiple-Hole Arrays of Composite Material Fasteners", Proc. Spring 1986 Conference on Experimental Mechanics, Soc. Exp. Mech. New Orleans, LA (June 1986)

[18] Bayer, M.H., Cloud, G.L., "Moiré Study of Strains Near Washers on Composites", Proc. Fall 1986 Conference on Experimental Mechanics, Soc. Exp. Mech., Keystone, CO (June 1986)

[19] Herrera-Franco, P.J., Cloud, G., Zimmerman, K., "Moiré Study of 2- and 3-Hole Fastener Arrays in FGRP Composite", Proc. 2nd Conf. on Advanced Composites, American Soc. Metals and Eng. Soc. Detroit, Dearborn, MI (Nov. 1986)

[20] Procht, M.M. and Hill, H.N., "Stress Concentration Factors Around a Circular Hole in a Plate Loaded Through Pin in the Hole", J. Applied Mechanics, Vol. 7, March 1940, pp. 5-9

4 Advanced structural measurement techniques for advanced turbine engines

W.A.Stange

SUMMARY

This chapter presents an overview of current research efforts aimed at improving turbine engine structural instrumentation capabilities to meet the requirements of a Project Forecast II initiative. Emphasis is placed on nonintrusive concepts which will be applicable to the advanced engines currently being designed for initial operational testing around the turn of the century. Technologies to be described include: blade tip deflection sensors to determine dynamic stress, thermographic phosphors to measure metal temperature, double core fibre optics to measure both temperature and steady-state strain on metal surfaces and while embedded in composites, and magnetic domain steady-state strain sensors. Advantages of the various systems, potential problems and limitations, as well as an assessment as to when the various technologies are likely to be ready for actual engine test usage are discussed.

INTRODUCTION

An area of ever increasing concern and difficulty in the turbine engine community is the accurate determination of the strains and temperatures under which engine components must operate. As we strive for the high thrust/weight ratios demanded by the High Performance Turbine Engine Technology (HPTET) initiative, structural efficiency becomes even more important. To ensure components are not overdesigned and, are therefore heavier than necessary, it is vital that systems and instrumentation are developed which predict and subsequently confirm exactly what temperatures and stresses are encountered. Furthermore, some of today's engine test programmes are already limited by problems associated with current structural instrumentation. The state-of-the-art in structural instrumentation has many shortcomings in both strain gauges and thermocouples. Current turbine engine tests are particularly impaired by the fact that present instrumentation is commonly temperature-limited, short-lived, inaccurate, and either protrudes into the gas flow stream or requires trenching the structural component in order to embed the sensor.

Structural instrumentation is used to characterize the engine environment and to identify any potential problem which may occur during an engine test. Knowledge of a potential problem allows the test engineer to take corrective action before damage to, or failure of the engine occurs. The turbine engine industry also relies on structural instrumentation to verify structural design techniques and analytical prediction capabilities. These techniques are used to assess turbine engine component life and to predict strain and temperature at different operating conditions. The operating environment of today's turbine engine taxes every capability of current strain gauges and thermocouples. Consequently, we cannot adequately and reliably instrument current technology gas turbine engines, and for future engines the problem will grow even worse.

The emphasis which Project Forecast II (PF II) has placed on the HPTET initiative, presents instrumentation engineers with an enormous challenge. HPTET's goal of doubling propulsion capability by the year 2000, without reducing engine life or reliability, will necessitate radical changes in virtually all aspects of turbine engine technology. Achieving this goal will require tomorrow's engines to run at much higher temperatures while using virtually no cooling air, operate at even higher tip speeds, utilize an array of new materials, and incorporate radical changes in design philosophy. Each of these requirements hinders our ability to make adequate strain and temperature measurements on the various structural components.

Higher gas temperatures in conjunction with reduced cooling air result in radically higher component temperatures. This drastic increase in component temperature will make it more difficult to secure any contacting sensor to the component, cause increased sensor susceptibility to corrosion, and will necessitate even higher temperature capability than is currently available. Higher tip speed will cause higher levels of centrifugal stress, thus making sensor attachment more difficult, as well as making strain measurements more important. New materials pose problems which are distinct to the material being incorporated. New metal alloys pose potential problems with sensor bonding. Composite materials, because of their anisotropic character will require subsurface strain and temperature measurements, thus necessitating sensors which can be embedded in the composite without radically altering the material matrix. Ceramic materials, because of their lower emissivities (and higher levels of reflected radiation) are likely to preclude the use of optical pyrometry. Additionally, because of their high stiffness, ceramic materials will require strain sensors with gauge factors greater than that which is currently required. Totally new designs will, because of the uncertainty associated with analytical prediction techniques, require measurements to be made at a myriad of locations. The aforementioned problems, in conjunction with the need to preserve the actual structural and gas flow characteristics, have resulted in an unprecedented need for new and radically improved instrumentation.

In an attempt to satisfy the requirements for improved instrumentation capability, a number of research programmes are being conducted to investigate new sensors and techniques. A premise common to each of these efforts is that they emphasize the investigation of new or "revolutionary" concepts, rather than attempting to improve or evolve existing techniques. The overall goal of these research programmes is the development and demonstration of advanced sensors and systems which will be capable of accurately and reliably measuring strain and metal temperature in environments which will make them applicable to the engines of the future. To this end, the following techniques are currently being investigated.

BLADE TIP DEFLECTION SENSORS

This measurement system will use time measurements to determine the dynamic tip deflection characteristics of a stage on a running turbine engine. It will consist of a small number of probes mounted to the case of an engine directly outside a stage of interest (Figure 4.1) and computational and software systems to process the sensor output.

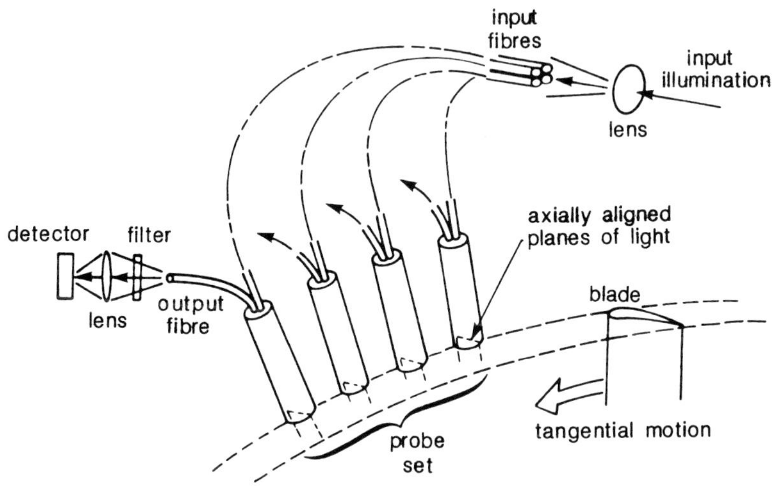

Figure 4.1: Non-Contacting Tip Deflection Measurement System

The blade tip deflection sensor is a very attractive alternative to conventional dynamic strain gauges. A major advantage of this technique is the fact that it is a noncontacting measurement system. As such, it requires no telemetry or slip rings to transmit the data, and has nothing affixed to the blade to disturb the flow field or other blade parameters.

A number of sources have done developmental work on various blade tip deflection sensors [1, 2, 3]. Previous engine test programmes have demonstrated the capability to measure nonintegral order vibration of fan and compressor stages. Work is on-going in this area to overcome the two major limitations which exist, i.e., the measurement of integral order vibrations, and measurement of turbine blade vibrations. Other less serious problems being addressed are attempts to reduce the number of sensors per stage, increasing the number of stages which can be monitored at a single time, and improving the accuracies of the measurements of amplitude and frequency.

The sensors will consist basically of two fibre optic bundles, a lens, and a photodetector. The input fibre optic bundle will carry light from a source and emit it through the lens and the case "window". The lens serves two functions, first, to focus the emitted light to the plane of the blade, and then to focus that light which is diffusely scattered by the blade tip's passage into the output fibre bundle. The output fibre optics then transmit the light pulse to the photodetector which emits an electrical pulse denoting the exact time of blade passage. This measured time of blade tip passage is then compared with the anticipated or expected time of blade tip passage to determine the time deviation. Tip deflection is then determined by the product of the time deviation and the circumferential velocity of the blade tip. In addition to providing an accurate indication of the instant of blade tip passage, the sensors will have to withstand high levels of temperature, pressure, and case vibration.

The output from the sensors is conditioned, digitized and organized into a data base. This data base is then mathematically converted into useful vibrational engineering data by means of a previously determined finite element library of modal characteristics. The software which will accomplish this task will basically do a fourier analysis of the frequency and amplitude content in the data base, thus establishing an inventory of the participating resonant modes, and the tip deflection amplitude associated with each one. The finite element library will provide a stress distribution for each mode/amplitude, and the integration of these stress patterns will yield the overall stress distribution of the blade.

The overall goal for this noninterference stress measurement system is a device which seems almost too good to be true, i.e., a system which measures stress for every airfoil on a stage over the entire surface of the airfoil with sensors easily removed from the exterior of the case in the event of sensor failure. This goal presents a number of challenges in several different technology areas, but the potential payoff, and the encouraging results encountered thus far make them well worth undertaking.

THERMOGRAPHIC PHOSPHORS

Turbine engine metal temperatures are currently measured using either thermocouples or optical pyrometry. Problems associated with these techniques limit both their accuracy and areas of applicability. Optical pyrometry is ineffective in those regions of the engine where either a high level of reflected radiation is present, or where the metal temperature is below approximately 1300°F. High levels of inaccuracy occur as the temperature approaches the lower bound of approximately 1300°F and when moderate levels of reflected radiation are present. Thermocouples are expensive, short-lived, require trenching when applied to blades or vanes, and necessitate telemetry or slip rings when applied to rotating hardware. Thermographic phosphors have the potential to alleviate many of the problems associated with current metal temperature measurement.

Thermographic phosphors exhibit a fluorescence spectrum whose spectral line amplitudes and decay times vary uniquely with temperature ([4] and Figure 4.2). These phosphors are typically excited to fluorescence by means of short wavelength (ultraviolet) pulsed laser radiation. This laser induced fluorescence (LIF) is then passed through an optical bandpass filter to select the desired emission line, and to eliminate any extraneous background radiation. The filtered LIF is then fed into a waveform-processing oscilloscope which determines the decay time of the fluorescent signal.

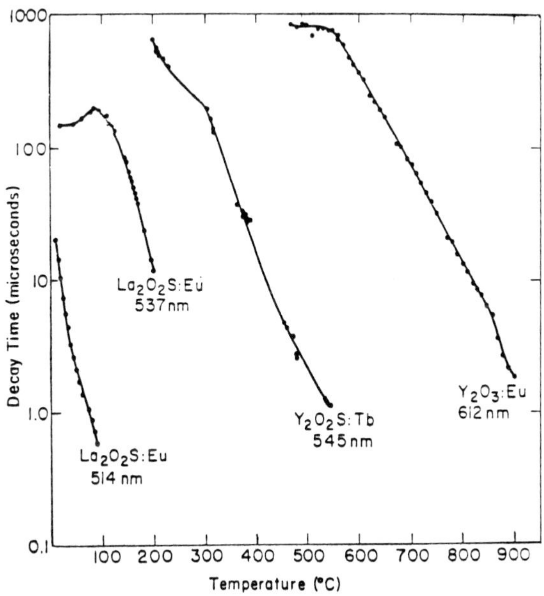

Figure 4.2: Fluorescent Decay Time vs. Temperature

Research previously conducted with this technique by the US Department of Energy has demonstrated the capability of certain thermographic phosphors to measure temperatures up to 500°F with very high levels of accuracy while bonded within an operating uranium enrichment centrifuge. The possible application of such phosphors to the rotating environment of a turbine engine is conceptually similar to the technique previously demonstrated on centrifuge rotors, the main difference being the temperature range to be measured. A basic research effort sponsored by the Aero Propulsion Laboratory was conducted to formulate and assess phosphors capable of measuring temperatures typical of those found in the hot section of a turbine engine. The results of that effort indicate that at least one and probably several phosphors exist which are capable of measuring temperatures to 2200°F or higher. Areas being emphasized under the current research programme include investigation and characterization of phosphors useful over wider and/or higher temperature range, and the development of phosphor bonding techniques applicable to advanced engine materials and operating conditions [5]. Engine test validation of this measurement technique should occur in 1990.

DOUBLE CORE FIBRE OPTIC STRAIN AND TEMPERATURE SENSOR

The Double Core Fibre Optic (DCFO) Strain and Temperature Sensor was initially envisioned as a surface mounted sensor which would provide strain information at temperatures beyond what could be achieved with strain gauges, and with minimal surface/flowpath interference [6]. As this effort progressed however, it became obvious that this sensor could provide temperature as well as strain measurement capability. Additionally and most importantly, the DCFO appears to be capable of determining subsurface strain and temperature in composite materials.

Fibre optics transmit light through a transparent core with the confining mechanism provided by the total internal reflectance at the interface between the core of high refractive index n, and the surrounding or cladding material of lower index of refraction n_2 ($n_2 < n_1$).

An optical waveguide formed from an array of two closely spaced, light-trapping cores in a common cladding (Figure 4.3) will exhibit a phenomenon known as "cross talk" when the optical fields in one core are coupled to the adjacent cores (Figure 4.4). This occurs because light energy is not totally confined within a core but, in fact, extends to a small degree into the cladding and the adjacent core. If one of the cores is illuminated then light will couple or cross-talk between the adjacent core. Any elastic or thermal perturbation will, to some extent (depending on the wavelength of light), influence the rate at which cross-talk occurs along the length of the fibre.

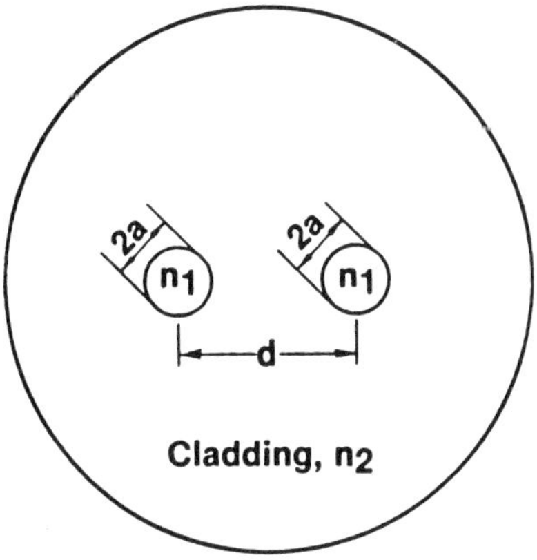

Figure 4.3: Cross Section of Double Core Fibre Optic

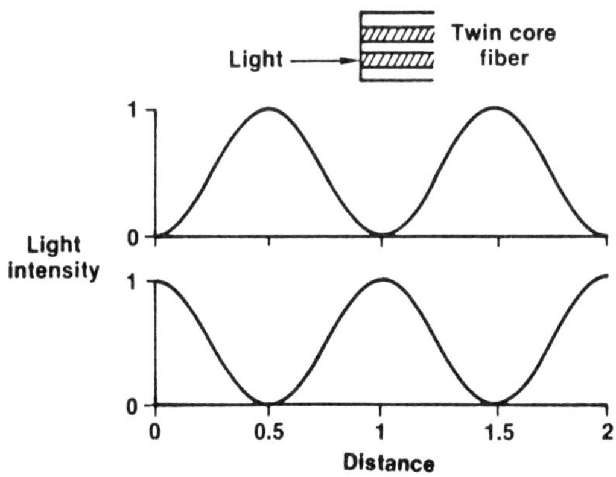

Figure 4.4: Cross Talk in a Double Core Fibre

By carefully controlling the design parameters of the fibre, (i.e. indices of refraction of the cores and cladding, distance between cores, and diameter of the cores) the "beat length", or distance required for a complete transference of light from one core to the other can be controlled for the wavelength(s) of input light to be used.

A double core fibre which is to be utilized solely for sensing strain will be designed such that, at the wavelength of input light, the fibre will be temperature-compensating(or temperature-independent). In this case the fibre would simply be calibrated over the strain range of interest without regard to temperature effects.

The approach taken by this research effort to extend the DCFO's capability to that of a dual function sensor involves using two wavelengths of input light. In this case the fibre would be temperature dependent for at least one of the input wavelengths, and a different "beat temperature" (i.e. the temperature change which will cause a full cycle of cross talk) will exist for each of the two wavelengths. By knowing the rates at which cross talk varies with strain and temperature at the two wavelengths [7], a unique combination of strain and temperature will exist for each pair of cross talk measurements (Figure 4.5).

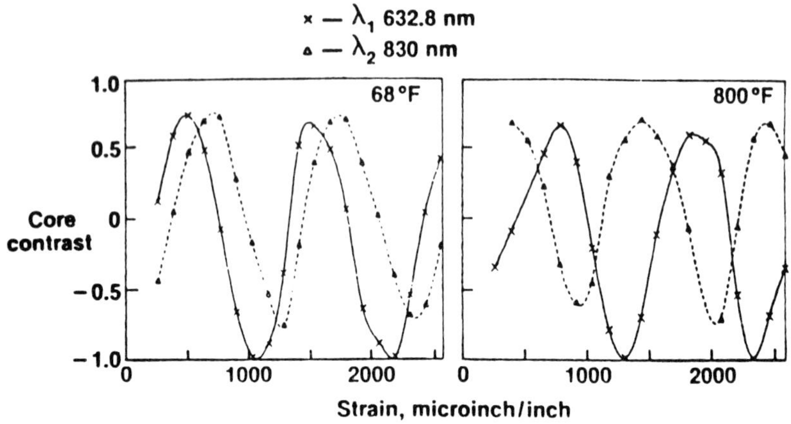

Figure 4.5: *Experimental Results Obtained Over a Range of Strain Levels at Different Temperatures*

The area in which the DCFO, and fibre optic sensors in general, have the potential to significantly improve measurement capability is the determination of subsurface strain and temperature in composite materials. Composite materials offer mechanical designers the potential of equal or superior strength and fatigue properties, at densities and weights which are significantly less than their metallic counterparts. These benefits are possible because composite materials are inhomogeneous and nonisotropic. These properties, although highly beneficial from a structures point of view, are a nightmare as far as structural diagnostics are concerned. The basic problem is that composite structures can have drastic nonlinearities in the amount of stress which they are able to accommodate without damage. Composite structures can also have drastic nonlinearities in the way in which stresses are distributed throughout the structure. It is because of these nonlinearities that surface-level strain and temperature measurements are not adequate to determine critical load limits for a structure. A fibre optic sensor embedded in the matrix of a composite would be a source of little, if any, disruption to the internal structural matrix, while providing valuable subsurface data on the nonlinearities inherent to the structure. A research effort has recently been initiated to evaluate a variety of fibre optic sensors throughout a diverse range of composite material compositions. This effort, which will be completed late in 1990, will provide valuable information on embedding fibre optic sensors in a number of different composite matrices.

MAGNETIC DOMAIN STRAIN SENSOR

The state-of-the-art for measuring steady-state strain during experimental engine tests is the resistive strain gauge. These gauges, first developed in 1938, make use of the fact that certain materials exhibit a change in electrical resistance which can be correlated with strain. Because current strain gauge alloys exhibit erratic resistance changes above approximately 750°F, steady-state strain measurements are limited to that temperature. Numerous research efforts conducted over the past decade have attempted to develop a high temperature resistive strain gauge alloy. The result has been the development of alloys which have measured strain to 1200°F in a laboratory environment. The requirements of HPTET however preclude evolving alloys to improve temperature capability at this kind of pace.

The magnetic domain strain sensor programme will investigate the relationship which exists between static strain and the magnetic properties of certain ferromagnetic materials. By utilizing the magnetostriction phenomenon and/or the Barkhausen effect, it is hoped that a sensor can be developed which will be more accurate and have higher temperature capability than today's electrical resistance sensors. Preliminary work in this area indicates that magnetic pulse transient time is a function of strain (Figure 4.6) and that this relationship is a well behaved function up to virtually the melting point of the material. A sensor based on these characteristics would be intrinsically more accurate, as the level of strain would be determined by magnetic pulse transient time measurements, which are more easily and accurately measured and have a much better signal-to-noise ratio than electrical resistance measurements. Additionally, these measurements should be attainable at temperatures approaching the melting point of the gauge, a characteristic which should greatly extend the temperature range over which strain measurements can be made.

This effort, which is in a very preliminary stage, is investigating basic phenomena associated with the sensor, and investigating the feasibility of using thin film technology to make the gauge nonintrusive.

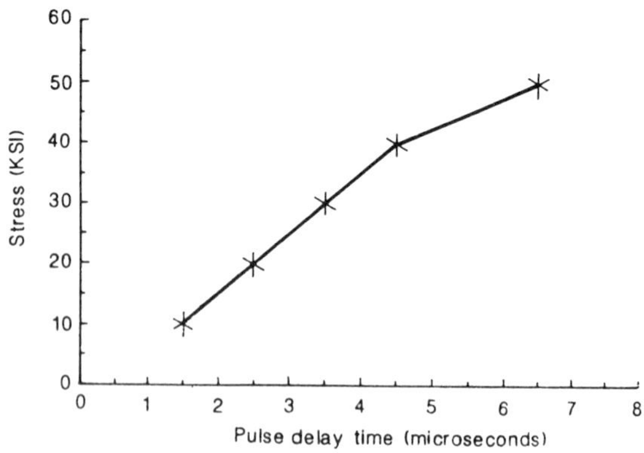

Figure 4.6: *Delay in Magnetic Pulse Propogation Time as a Function of Stress*

PROMISING CONCEPTS REQUIRING FURTHER BASIC RESEARCH

A number of techniques have, at least in theory, the potential of solving virtually all of the problems associated with adequately determining the sorely needed structural and environmental parameters for current and future turbine engines.

Basic research efforts are vital to assessing the following concepts in order to evaluate their potential benefits. Neutron diffraction is a totally noninvasive nuclear based concept which has the potential of allowing strain to be measured internal to the engine. Related nonintrusive nuclear techniques which have the potential of providing remote temperature sensing include resonant neutron thermometry and photoexcitation of resonant nuclear levels by gamma radiation. Additionally, a concept known as the superlattice strain gauge has potential as a possible means of optically determining strain at elevated temperature. This sensor, as envisioned, would be based on layered epitaxial semiconductor crystal-structure technology.

CONCLUSIONS

The state-of-the-art in structural instrumentation does not adequately address the needs of today's turbine engine test programmes. The HPTET initiative with its higher temperature and new materials requires dramatic increases in instrumentation capability. Although a number of efforts are underway to address this need, additional research programmes are imperative to providing the advances which are required.

REFERENCES

[1] Roth, H., "Vibration Measurements on Turbomachine Rotor Blades with Optical Probes", Measurement Methods in Rotating Components of Turbo-Machinery, pp. 215-224, March 1980

[2] Watkins, W.B., Robinson, W.W., and Chi, R.M., "Noncontact Engine Blade Vibration Measurements and Analysis", AIAA paper 85-1473, AIAA/SAE/ASME/ASEE 21st Joint Propulsion Conference, July 1985, Monterey CA

[3] Jones, H.T., "Development of a Noninterference Technique for Measuring Turbine Engine Rotor Blade Stresses", AIAA paper 85-1472, AIAA/SAE/ASME/ASEE Joint Propulsion Conference, July 1985, Monterey CA

[4] Noel, B.W., et al, "A Proposed Method for Remote Thermometry in Turbine Engines", AIAA paper 85-1468, AIAA/SAE/ASME/ASEE 21st Joint Propulsion Conference, July 1985, Monterey CA

[5] Borella, H.M., Franks, L.A., and Noel, B.W., "Development of Bonding Techniques for High-Temperature Applications of Thermographic Phosphors", Society for Experimental Mechanics, Third Annual Hostile Environments and High-Temperature Measurements Conference, March 1986, Cincinnati OH

[6] Dunphy, J.R., and Meltz, G., "Development of a Fiber Optic Sensor for Turbine Disk Diagnostics", AIAA paper 85-1469, AIAA/SAE/ASME/ASEE 21st Joint Propulsion Conference, July 1985, Monterey CA

[7] Meltz, G. and Dunphy, J.R., "Twin-Core Fiber-Optic Strain and Temperature Sensor", OSA/IEEE Third International Conference on Optical Fiber Sensors, February 1985, San Diego CA

5 Transducers for the measurement of surface strains and global deformations of tyres under driving conditions

J.Neugebauer & V.Grubisic

SUMMARY

The deformation behaviour of tyres is an important quantity with respect to tyre performance, vehicle handling and especially tyre durability. In order to obtain deformation data from tyres under realistic driving conditions appropriate deformation transducers had to be developed. Within this chapter a transducer capable of measuring high surface strains and a global deformation transducer, measuring the vertical and lateral displacement of the tyre's crown is being presented. Some applications will be demonstrated.

INTRODUCTION

The transducers described in this chapter provide an additional tool to be used for the development of tyres. Information gained assists in the design phase when the tyre is computer-modelled, as well as in the test phase when the various properties of a tyre are being tested. During the design phase intensive analytical work, increasingly based on the finite element method is carried out by the tyre manufacturer. Since the tyre is a rather complex design object consisting of a number of different rubber components as well as reinforcing cords of various materials and geometrical configuration, the computer modelling technique is quite complicated and involves high cost and a considerable amount of time.

When a tyre has been manufactured a number of tests are carried out in order to assure the quality of the product. Besides the verification of the performance data carrying out non destructive tests like the uniformity test also the durability properties like wear, structural fatigue, and strength are investigated by subjecting tyres to destructive tests. Transducers measuring tyre deformation provide several advantages:

- Experimental deformation data under real service conditions permit an improvement and verification of present and future analytical models.
- The measurement of deformation data under typical laboratory

loading conditions allows the quantitative comparison of various design concepts.
- The measurement of deformation in service allows better description of those physical quantities which are being held responsible for the structural fatigue of the tyre's carcass.

The latter point will be of increasing importance especially for truck tyres which will be retreaded more often in the future for economic reasons. Two different types of tyre deformation are being treated in this chapter:
- local deformation measured at the surface of the tyre, referred to as "local surface strain",
- global deformation of the crown of the tyre with respect to the rim measured in the vertical and lateral direction referred to as "global vertical and lateral crown displacement".

The following sections contain a description of the transducers and some applications.

MEASUREMENT OF SURFACE STRAINS

Problem

The problem of strain measurements of highly deformable components has been addressed before [1-4]. Different approaches ranging from standard and high strain foil gauges to elongation transducers to be inserted into the tyre are known. The principle of liquid metal gauges could not be applied successfully up to now, due to the lack of linearity or reproducibility of results. Besides the demand of a linear strain vs. signal relation the transducer itself may not constrain the deformation of the component to be measured.

Rubber components are designed and used to endure and transmit large deformations, thus involving large surface strains. Besides tyres there are quite a number of components which are subjected to large deformations such as rubber hoses, engine supports, compensators, tube connectors and membranes, reinforced rubber belts, and many others. Local strain distributions may show considerable gradients (strain concentrations) along the surface of the component. Since most transducers have an integrating effect over their individual measuring length it is important to provide transducers with different effective measuring lengths. In addition to the capability of measuring large local strains an appropriate transducer must also be designed to withstand the outdoor testing environment.

Description of the transducer

Figure 5.1 shows two transducers [5] of different sizes (13 x 16mm and 6 x 11mm). The transducer consists of a rubber body into which a rubber tube is embedded. The tube is wound back and forth several times in order to increase it's total length.

Figure 5.1: *Transducers for the measurement of large surface strains (liquid metal gauges) [5]*

After the fabrication of the transducer the internal tube is filled with mercury and two leads are inserted into the open ends to provide the electrical connection. The measuring principle is basically that of strain gauges. A change in the length of the transducer results in a change of the resistance of the mercury filled tube, measurable at the two leads. This change of resistance may be measured using a Wheatstone Bridge circuit. Since the resistance of the transducer ranges between 0.2 and 1.0Ω (depending on size) an additional resistor of 120Ω is integrated into circuit. Thus the signal can be processed using standard strain gauge equipment. The transducers are bonded to the measuring object using a special adhesive.

Properties

Calibration
One of the outstanding properties of the developed transducer is its linear relationship of strain vs. measuring signal (Figure 5.2) where the calibration characteristics of 6 gauges are shown, which were placed on a smooth rubber specimen. Subjecting the rubber specimen to an axial load and externally measuring the relative elongation equivalent to the local strain, the given relation of strain vs. measuring signal was obtained. The shaded area indicates a scatter band of ±10% for the six transducers. The gauge factor k_G was determined to be 0.0118, i.e. about 170 times less than the gauge factor of a common strain gauge ($k_G \approx 2$). The linear measuring range of these liquid metal gauges is greater than ±20%.

Temperature sensitivity
Since it is known that mercury is sensitive to temperature changes, special tests were conducted using the same rubber specimen and test procedure as above to produce a calibration characteristic under elevated temperatures. Figure 5.2 shows for temperatures of 50 to 60° and 80°C that the slope of the characteristic

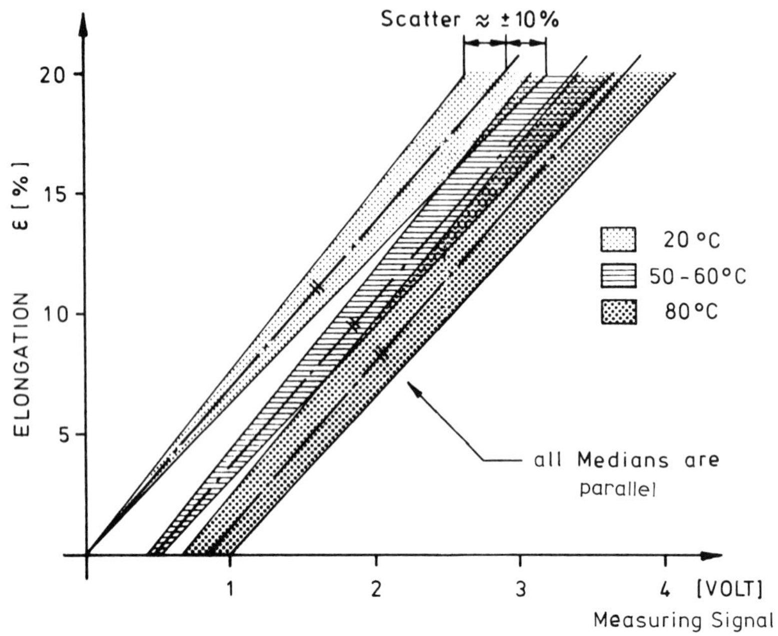

Figure 5.2: *Calibration characteristic of liquid metal gauges at room temperature and at elevated temperatures*

does not change, i.e. that the sensitivity of the gauge is not influenced by temperature. However, a certain amount of drift of the characteristics is observed. Additionally the different gauges react a little differently with temperature, which tends to spread the scatter band. For practical applications where heating-up of rubber components has to be expected (e.g. with engine supports or tyres), the increase of temperature will usually be fairly slow so that the drift can be eliminated by balancing the measuring circuit after certain periods of time.

Inertial effects

If the transducer is applied in a dynamic test where large centrifugal effects occur, certain limitations to the use of the transducer must be considered. These limitations refer mainly to the point of application and the orientation of the transducer with respect to the direction of the centrifugal force. In practical applications, however, these liquid metal gauges proved to be successful when applied to passenger car tyres up to speeds of about 60 to 70 km/h. From experience it is known that this range of velocities is quite sufficient to carry out extreme cornering tests on skid pads and straight driving tests on rough roads (cobblestone and pothole tracks).

Endurance

Endurance tests of the bonding between the transducer and tyre were carried out under strain levels of about 10%. After about 2.10^6 cycles (equivalent to approximately 4000km driving with a passenger car tyre) no debonding was noted.

Another advantage of the transducer became evident when a wire lead happened to be pulled out of the tube or when measurements had to be repeated after a long period (during which the object under measurement was stored away). In this case the transducer may be "repaired" by refilling it with mercury after which it is ready for measurements again.

Application

The transducer was primarily developed for tyre research and development. Figure 5.3 shows the application of gauges on the inside and outside of a tyre. A complete strain distribution along the inside contour of this tyre under various loading conditions simulated in the laboratory is given in Figure 5.4. These distributions may be used to verify and improve computer modelling technique (results of the FE-analysis).

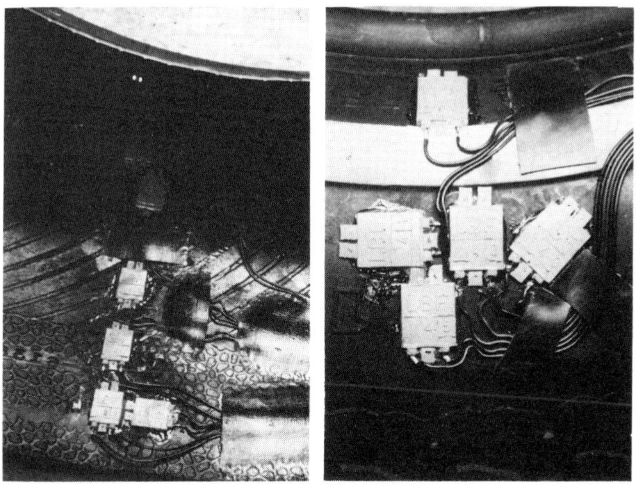

Figure 5.3: Application of liquid metal gauges on the inside and outside contour of a passenger car tyre

It is also possible to apply the transducers under different angles to form a rosette (Figure 5.3) allowing the calculation of principal strains and their directions. The transducer was mainly designed to be used under operational conditions, e.g.

driving on a proving ground or on public roads. An example of the strain recording obtained from a liquid metal gauge positioned on the inside of the right front wheel tyre is shown in Figure 5.5. The top recording was taken from a pothole track at about 35 km/h while the two other recordings were obtained from extreme cornering manoeuvres.

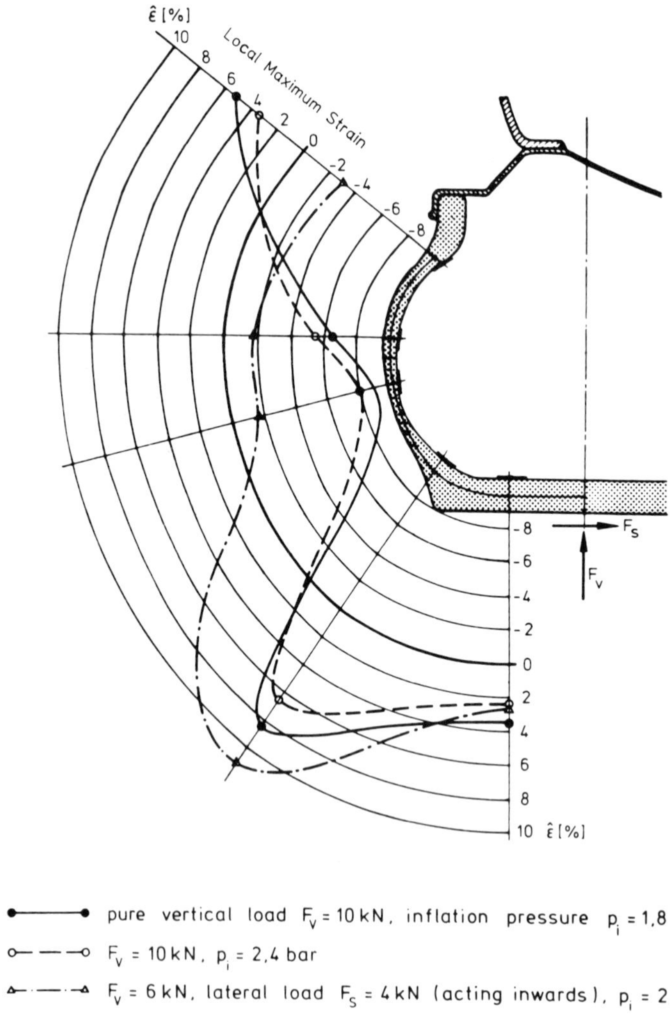

Figure 5.4: *Strain distribution on the inside of a tyre measured radially under various loading conditions*

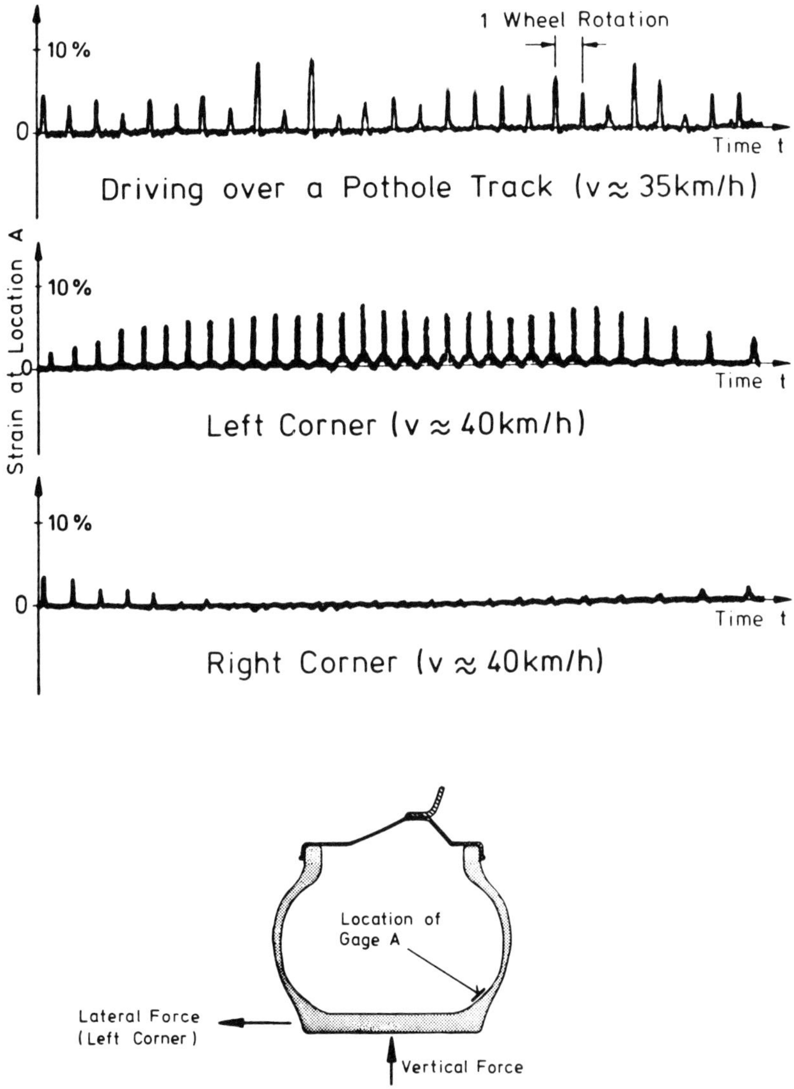

Figure 5.5: *Recorded strain data from a location on the inside of the tyre driving over a pothole track and cornering*

MEASUREMENT OF GLOBAL DEFORMATION

Problem

The objective of the development was to design and implement a transducer capable of measuring the global tyre deformations under driving conditions. Generally, the global deformations are defined as the vertical and lateral displacement of a point on the inside contour of the tyre with respect to the rim. The global tyre deformation has been measured in order to determine the vertical and lateral tyre stiffnesses. These quantities are important for the handling characteristics of the tyre and the dynamic wheel loads.

In the following a transducer [6] is described which is capable of measuring the relative movement of the inside centre point of the footprint, referred to as the crown displacement with respect to the bead area. The latter is attached rigidly enough to the rim to permit the influence of the small relative displacement of the bead and the rim to be neglected

Description of the Global Deformation Transducer

Figure 5.6a shows a cross section of a passenger car tyre and the installed global deformation transducer. It consists of two spring loaded measuring elements, each of which is attached to the inside centre point (CP) of the footprint and the bead (points P_1, P_2) using specially designed screws. The relation of springload vs. elongation Δl is linear. Therefore, the signal obtained from the two integrated load cells is a function of the elongation of each measuring element. Figure 5.6b shows the geometrical relations. Besides the undeformed transducer, which is described by a symmetrical triangle with the base B and the height H resulting in a length l_0 of each measuring element. Figure 5.6b shows the deformed transducer. This conversion of the measuring signals is carried out by analogue computer analysis which is extremely advantageous when on-line data processing is required during outdoor tests. While the base B remains constant the length of each measuring element changes from l_0 to l_1 or l_2 respectively. The differences Δl_1 and Δl_2 are converted into vertical (X_V) and lateral (X_L) global deformations through the two equations given in Figure 5.6b. Generally, the centre point CP has 6 degrees of freedom. Besides the two translatory movements X_V and X_L the CP can move circumferentially e.g. when large braking loads are applied. In that case the measurement may be influenced to quite an extent. In order to avoid this, an additional measuring element has to be incorporated thus forming a pyramid instead of a triangle. The rotational movements can usually be neglected since the ends of the measuring elements are free to rotate about two separate points extremely close to the CP.

(a)

(b)

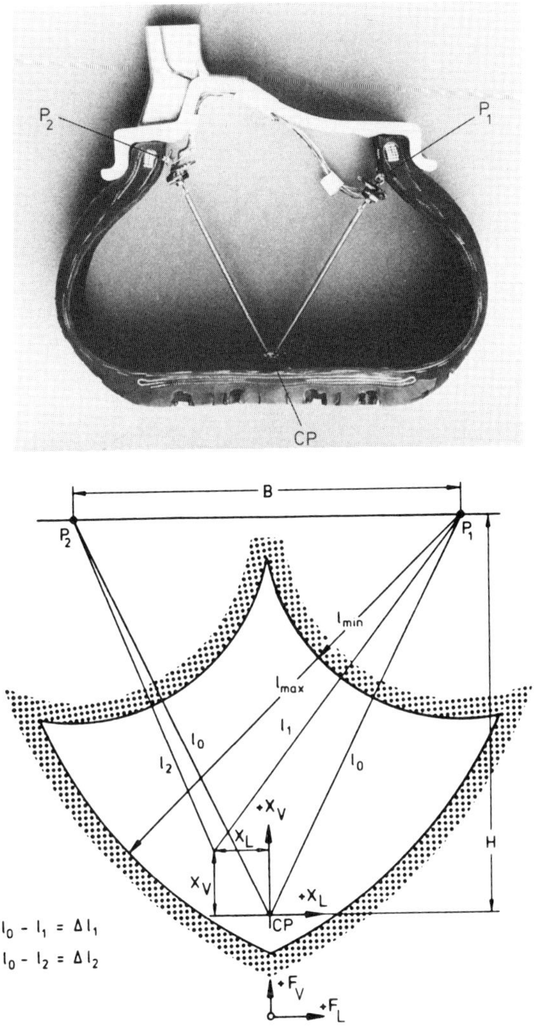

$$X_L = \frac{\Delta l_1^2 - \Delta l_2^2 + 2\, l_0 \cdot (\Delta l_1 - \Delta l_2)}{2\,B}$$

$$X_V = H - \sqrt{(\Delta l_2 + l_0)^2 - (\frac{B}{2} - X_L)^2}$$

Figure 5.6: *(a) Global displacement transducer installed into the cross section of a passenger car tyre [6]*
(b) Mathematical relations for the derivation of the global deformations from the measuring signals and measuring range

Properties

Measuring range

The measuring range of the global displacement transducer depends on a number of influences:

- length l_0
- type of spring
- required service life of the transducer
- required maximum speed
- required measuring accuracy.

The global deformation transducer shown in Figure 5.6a was designed for passenger car tyres. The length l_0 is 130mm and the measuring range indicated by the shaded area in Figure 5.6b is limited to a maximum length of 140mm and a minimum length of 60mm thus resulting in a possible difference of Δl = 80mm. Depending on the above mentioned requirements it is possible to produce global deformation transducers for different applications.

Velocity

The velocity under which measurements are of importance may be divided into three ranges: low velocity up to about 25 km/h, medium velocity up to about 70 km/h, high velocity from 70 to 140 km/h (possibly above). The first range is used in quasistatic tyre analysis. The transducer may be used as shown in Figure 5.6a. For the second range, which represents common velocities for outdoor tests, e.g. driving on skid pads or rough roads (pothole, cobblestone, etc.) the influence of centrifugal forces becomes evident. The transducers will provide a signal which increases linearily with the velocity thus resulting in a shift of the mean value of the recording. In order to avoid this, a special compensation device as shown in Figure 5.7 has been developed, which is bolted to the wheel. It consists basically of two elements which linearily react towards centrifugal forces. Their signal is fed into the electrical circuit of the actual transducer thus eliminating the influence of centrifugal forces. The signal is transferred through a slip-ring transducer to the analogue computer. The third range of velocities about 70 km/h are mainly used in long term endurance tests and high speed performances test on the proving ground. Besides the necessity of installing the compensation device one has to observe the natural frequency of the transducer which is closely connected with the measuring range and the influencing quantities. For the shown design the natural frequency was about 160 Hz.

Special filtering techniques are capable of widely eliminating the influence of the natural frequency, unless it is contained in the actual signal to be measured. In those cases special light weight, high frequency designs may be used.

Application

Figure 5.8 shows a recording from a proving ground test. The first two tracks give the vertical and lateral crown displacement when driving over a pothole track at about 35 km/h. Each peak

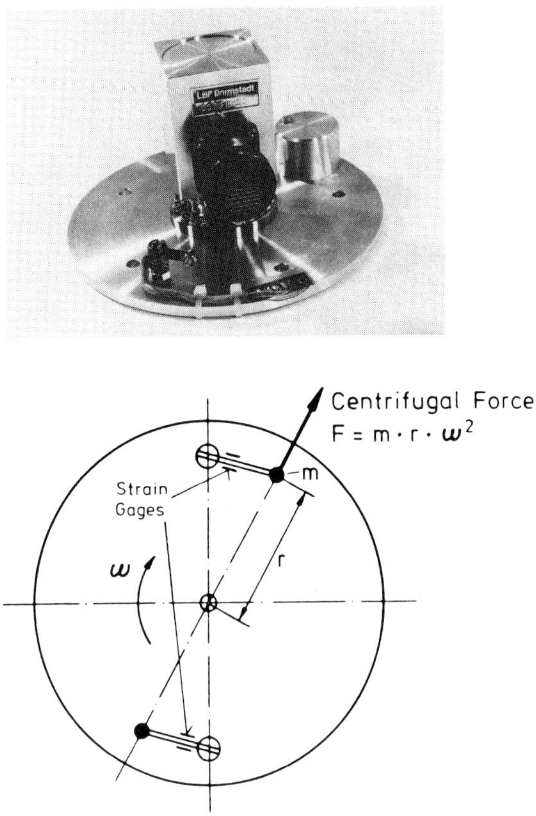

Figure 5.7: *Compensation device for eliminating the influence of centrifugal forces on the global displacement transducer*

in the top recording represents one wheel rotation. The second two recordings were obtained on a skid pad turning left with the instrumented tyre on the right front position thus increasing the vertical load and building up lateral forces acting towards the centre of the vehicle, thus being positive by definition (see also Figure 5.6b). The last two recordings were obtained while turning right. The vertical crown displacement decreases as does the vertical load while the lateral displacement becomes negative, equivalent to a fairly small lateral load acting towards the outside of the vehicle.

To discuss all possible applications of the global deformation transducer would exceed by far the scope of this chapter. However, it ought to be mentioned that it is also possible to measure the global deformation of other points along the inside contour of the tyre and not be restricted to the centre point only.

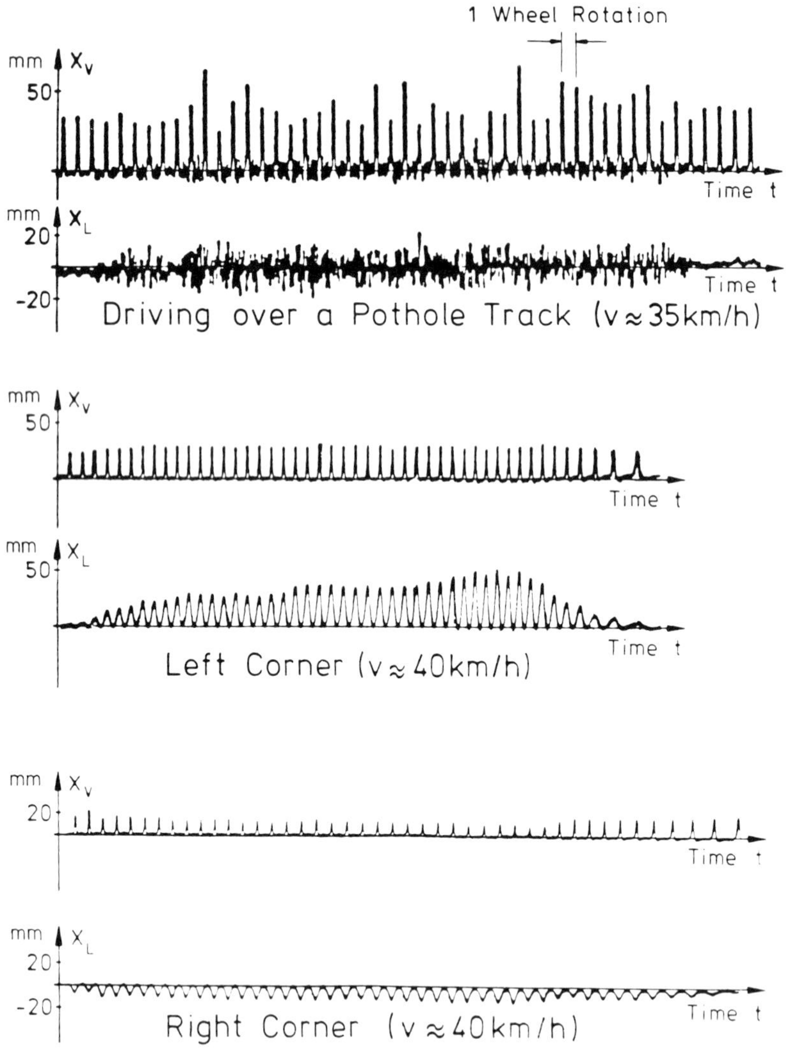

Figure 5.8: Recorded global vertical and lateral deformation data driving over a pothole track and cornering

Another important application could consist in adding another measuring element in order to obtain global deformation data in the circumferential direction, e.g. due to braking.

REFERENCES

[1] Kern, W.F., Meßelemente für die Anwendung auf Gummiprodukten; Vortrag VDI-Seminar "Spannungsanalyse mit Dehnungsmeßstreifen" Karlsruhe (1985)

[2] Kern, W.F., Verformungs- und Spannungsmessung an Gürtelreifen, Teil I, II; Kautschuk und Gummi, Kunststaffe, 24 (1971), Heft 11, S. 599-608, und Heft 12, S. 666-669

[3] Hurry, J.A., Woolley, R.P., A New High Range Strain Gage, Rubber Age, Sept. 1953

[4] Clark, S.K., Dodge, R.N., "A Load Transducer for Tire Cord", SAE Paper 690 521 (1969)

[5] Meßwertaufnehmer Zur Erfassung hoher örtlicher Oberflächen-dehnuingen, Deutsches Patent P 3544 894.6-52

[6] Verfahren zur Messung der globalen Verformung eines Reifens, Deutsches Patent P 3544 893.8-52

6 Calibration of a strain gauge measurement system for use in modal analysis

K.G.McConnell & M.K.Abdelhamid

SUMMARY

A concern in modal analysis is the proper calibration of the strain gauge measurement system. Of primary concern is both the unit sensitivity and the actual time delay that exists between the input force transducer system and the response of the strain amplification system. An electronic calibration scheme can be used to establish the amplifier time delay. A free-free beam calibration has been proposed but experiments show that the free-free beam impulse response is not well enough defined to be usable for calibration purposes. An axial rod calibration with a known time delay is proposed as a reasonable alternative to obtain overall measurement system performance.

INTRODUCTION

In order to adequately perform modal analysis, it is necessary to obtain the frequency response function (FRF), $H_{pq}(w)$, between the output at p and due to the input at q. Thus,

$$H_{pq}(w) = \frac{a_p(w)}{F_q(w)} \tag{6.1}$$

where
$a_p(w)$ = frequency spectrum of structural response signal
$F_p(w)$ = frequency spectrum of input force signal

The measurement system produces two voltages, $E_p(t)$ for output response and $E_q(t)$ for input force. The frequency spectra of these voltages are

$$E_p(w) = H_p(w) S_p a_p(w) \tag{6.2}$$

for output response and

$$E_q(w) = H_q(w) S_q F_q(w) \tag{6.3}$$

for input force

where

$H_p(w)$ = FRF of the structural response measurement system

S_p = voltage sensitivity of the structural response measuring system (V/unit a)

$H_q(w)$ = FRF of the input force measurement system

S_q = voltage sensitivity of the input force measuring system (V/unit F)

The frequency analyzer produces

$$H_{pq}(w)_m = \frac{E_p(w)}{E_q(w)} = \frac{H_p(w)}{H_q(w)} \frac{S_p}{S_q} \frac{a_p(w)}{F_q(w)} \qquad (6.4)$$

which can be written as

$$H_{pq}(w)_m = H_{pq}(w)\, S_{pq}\, H_I(w) \qquad (6.5)$$

where

$H_{pq}(w)_m$ = measured FRF

$H_I(w)$ = $H_p(w)/H_q(w)$ = FRF of combined measurement system

S_{pq} = S_p/S_q = system sensitivity calibration constant

Ideally, $H_p(w) = H_q(w) = H_I(w) = 1.0$ so that only the system sensitivity S_{pq} need be known. In the following sections, three methods for measuring $H_I(w)$ are presented.

ELECTRONIC CIRCUIT CALIBRATION METHOD

To calibrate a given strain gauge system for dynamic measurement, a known time history should be applied to the strain gauge. This strain history should be well defined so that the desired calibration can be performed over the frequency range of interest.

Factors that influence the dynamic response of a strain gauge measurement system include the dynamic response of the strain bridge amplifier, the type of lead wires used to connect the strain gauge to the bridge amplifier, the dynamic response of the antialiasing filter, if used, and the analyzer's hardware and software. Magnitude ratio and phase shift frequency functions define the dynamic response of the measurement system, i.e. determine the strain system sensitivity S_p and frequency response function $H_p(w)$.

An experiment is presented here that uses an electronic circuit to estimate the dynamic calibration function $H_p(w)$. This electronic circuit is designed to shunt the strain gauge for a short period of time. Moreover, a method to eliminate most of the phase lag included in $H_p(w)$ is presented.

A strain gauge (Micro Measurement Group type EA-06-240LZ-120) was glued onto an aluminium surface using Permabond 910 adhesive. A 1.0 metre long twisted wire pair was soldered to the strain gauge and the other end was connected to the strain bridge. The two terminals of the electronic calibration circuit were connected to the strain gauge terminals. This electronic circuit was designed to apply a shunt resistor of 24.0 Kohms for a period of 50 microseconds. The output of the bridge amplifier was connected to the input of the analyzer and properly grounded. The electronic circuit was operated 50 times and the average FRF is shown in Figure 6.1(b). Figure 6.1(a) shows the actual signals analyzed. The averaging process is meant to cancel the uncorrelated noise that contaminates the input and output channels. The uncorrelated noise is generated by the digital quantization process, thermal noise in the electronics, and surrounding electromagnetic fields.

The input and output signals are shown in time domain in Figure 6.1(a). The rectangular input signal (measured voltage across the shunt resistor using high quality shielded cable) exhibits dynamic ringing. This ringing on the sharp corners of the signal is created by the input stage antialiasing filters of the analyzer (B & K model 2032). The frequency analyzer (NORLAND model 3001) does not have these filters and did not show this ringing in the time domain signal. This indicates that the shunt resistor applied to the strain gauge is actually a clean rectangular function with a rise time less than 5 microseconds as measured by the NORLAND. The output of the bridge amplifier exhibits round edges and a time delay relative to the input pulse. The system response can be evaluated by studying the FRF shown in Figure 6.1(b). The magnitude ratio stays in the range 0 to -5 dB in the frequency range of 0 to 40000 Hz and then drops by almost 10 dB in the next 4000 Hz. The phase plot drops with a linear trend at the rate of 0.0289 degrees/Hz (a drop of 115.4 degrees over 4000 Hz). The linear approximation of the phase shift is not valid for frequencies above 4000 Hz. Figure 6.1(b) gives a complete evaluation of the dynamic response of the entire strain measuring system.

The linear phase lag of the measuring system can be corrected for by shifting the output signal in time in order to cancel the time lag. This practical and simple method is now demonstrated using the strain measuring system under consideration. The phase lag slope was calculated from Figure 6.1(b) as follows:

$$S = \frac{D\theta}{Df} = \frac{115.4}{4000} = 0.0289 \text{ degrees/Hz} \qquad (6.6)$$

where

S is the phase slope in degrees/Hz
$D\theta$ is the phase shift over Df
Df is the frequency difference.

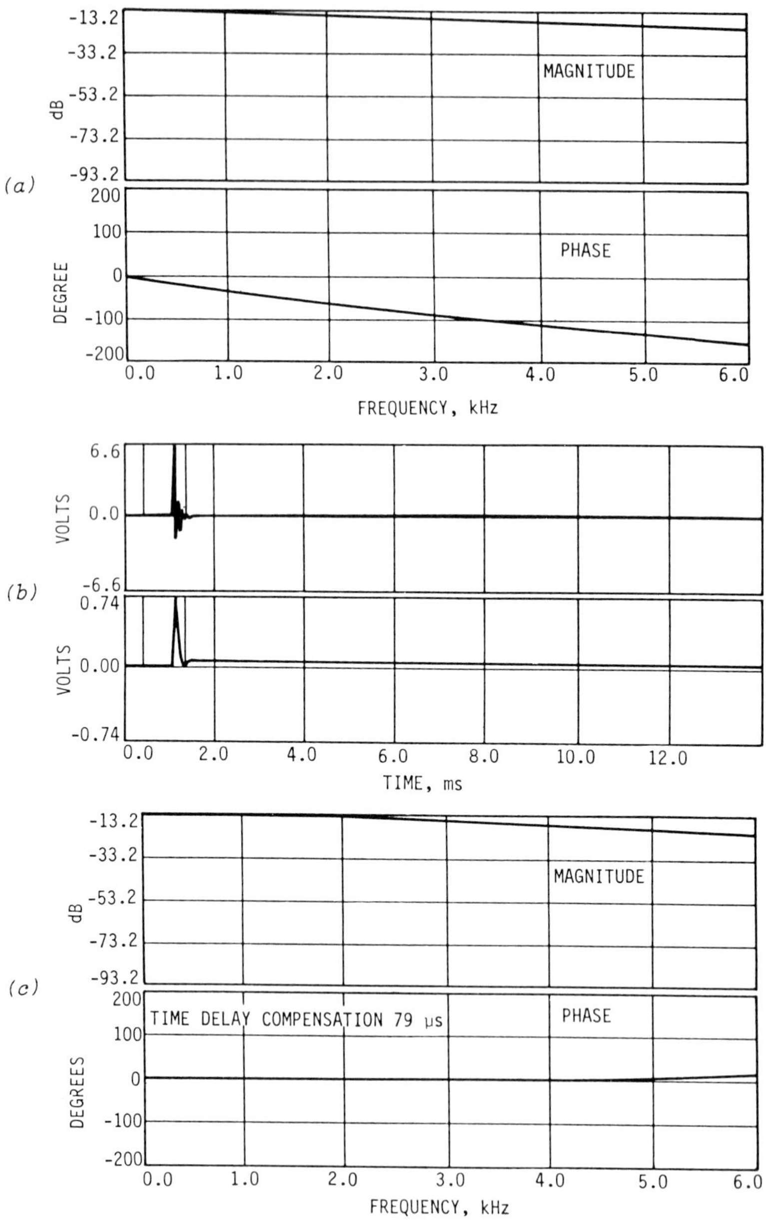

Figure 6.1: (a) FRF of the measurement system using electronic circuit calibration
(b) Electronic circuit input (top) and system output (bottom)
(c) FRF of the measurement system using electronic circuit calibration and applying 79 μs time delay compensation

The following formula is used to calculate the corresponding time compensation T

$$360\ T = 0.0289 \qquad (6.7)$$

The time shift compensation needed is T = 80 microseconds. This time shift when applied to the output turns the phase frequency range of 0 to 4000 Hz. A reversing amplifier will have a -180 degrees phase shift in its working region. Beyond the 0 to 4000 Hz working region, the amplitude as well as the phase shift differ from the sought after ideal of 0 dB and either 0 or -180 degrees. Thus, it is possible to improve the strain measuring system dynamics in the range of 0 to 4000 Hz by applying a time compensation in the frequency analyzer.

FREE-FREE BEAM CALIBRATION METHOD

In a paper presented at the Spring 1986 meeting of the Society for Experimental Mechanics [1], a short free-free beam was used in the dynamic calibration of a strain gauge measuring system. The validity of the conclusions arrived at in this paper was challenged in a later paper [2]. In the following, the results obtained from a similar short beam are presented and discussed as to why the views presented in the first paper [1] are incorrect, and how to correctly interpret that data.

A short steel beam with dimensions of 6.3 x 11.1 x 181.6mm was used for this study. The free-free fundamental natural frequency in lateral bending was 1000 Hz. A strain gauge was attached at the beam centre such that it will measure strains from those natural frequencies with mode shapes having no nodes at the beam centre. Theoretical analysis of free-free beams [3] shows that the first, third, fifth .. etc. natural frequencies satisfy this condition. The ratios of these frequencies to the first natural frequency are 1, 5.4, 13.3 .. etc. The free-free boundary condition was simulated by hanging the beam from two long strings. For simplicity, the free vibration was excited by impulsing the suspended beam near one end with an instrumented hammer in the horizontal direction; a technique similar to that used in the original paper [1].

Figure 6.2 shows the FRF for the 1000 Hz beam. The range of frequencies displayed is 0 to 1600 Hz. Large amplitude ratio is shown at the 1000 Hz resonance frequency. Other resonance frequencies that will be present in a larger frequency range is the third (5400 Hz), the fifth (13300 Hz) .. etc. The second natural frequency will be absent because the strain is measured at the beam mid-point which is the node point for the second natural frequency. The phase plot shows a nearly linear increase in phase lag in the region 0 to 1000 Hz. At the 1000 Hz a 180 degrees phase is added to resonance.

The slope of the phase plot in the region 0 to 1000 Hz is caused by the characteristics of the bridge amplifier. This

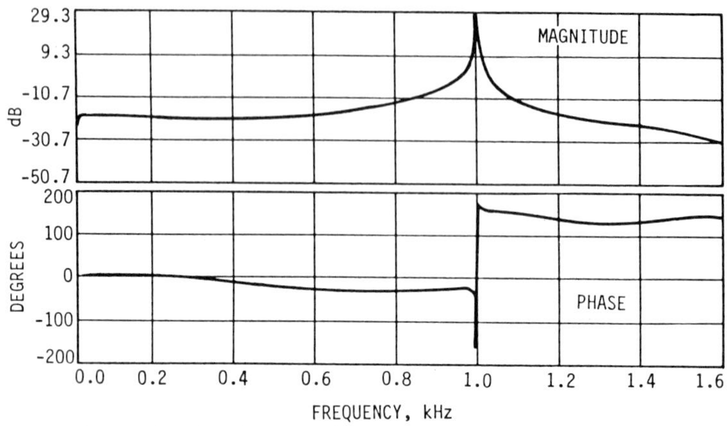

Figure 6.2: FRF of a laterally vibrating free-free beam

amplifier was tested before and was found to introduce phase shift of .0289 degrees per Hz in the region 0 to 4000 Hz. To compensate for this phase shift, an 80 microseconds compensation was used and the result is shown in Figure 6.3. The slope is eliminated in the phase plot in the 0 to 1000 Hz region. Hence, with the phase distortion of the amplifier compensated for, the phase plot shows the 180 degrees phase shift at the 1000 Hz natural frequency as expected.

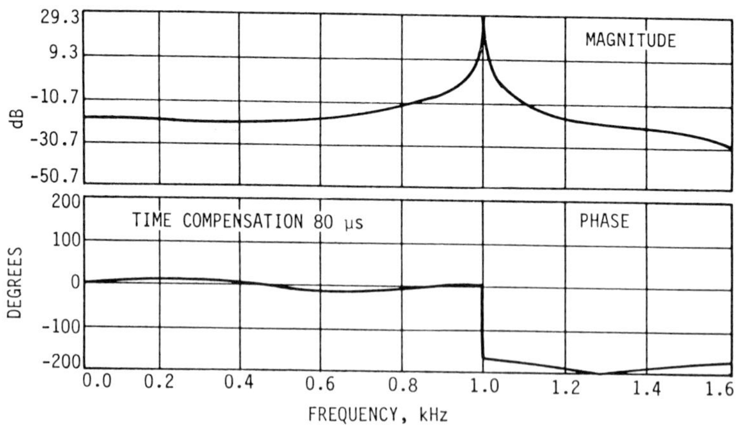

Figure 6.3: FRF of a laterally vibrating free-free beam with 80 microseconds delay time compensation

The free-free beam is difficult to use as a calibration system for several reasons. First, the force and strain are not simply related so no magnitude calibration can be obtained. Second, the beam must be very small to achieve higher frequencies. This small size leads to problems of exciting the structure at the same point. Third, it is difficult to extract the real time delay of the strain measurement system since each flexure waves travels with a different wave speed [5].

LONG BAR CALIBRATION METHOD

A well defined relation between the input force and the output strain is needed for calibration of a strain measurement system. Elastic wave propogation in bars satisfies this requirement. In the following, a method of using an impacted long bar for strain system calibration is presented.

The wave-form e(0,t) of the strain at the point of application of the impact force f(t) is related to the impact force by the following relation [3]:

$$f(t) = -EA\, e(0,t) \qquad (6.8)$$

Where E is the elastic modulus of the bar, and A is its cross sectional area. At distance x from the impacted end, the strain function is expressed by the following formula:

$$e(x,t) = -f(t - x/c) / EA \qquad (6.9)$$

Where c is the speed of elastic wave propagation in the bar. This speed is expressed by the following formula:

$$c = \sqrt{E/\rho} \qquad (6.10)$$

Where ρ is the mass density of the bar material.

Equation 6.9 is only valid in the first half cycle of strain history since the elastic wave reflects at the free end of the bar and returns as tensile wave after starting as a compressive wave. The amount of time available for equation (6.9) to be valid depends on the length of the bar and the duration of the elastic wave since these two waves can overlap if the bar is too short for a given pulse direction.

So, based on the aforementioned relations, it is clear that the strain history at distance x during the first passage of the elatic wave as it travels along the bar is a scaled replica of the force function that occurs x/c seconds earlier. An added time delay due to the dynamic characteristics of the strain measurement system affects the actual recorded strain.

A 2.57m long by 0.123m diameter aluminium bar was used for this experiment. The bar was suspended freely by two long strings. A thin steel anvil was screwed to the end of the bar, and a strain gauge was glued at distance 1.29m from the steel anvil surface. The strain gauge length was chosen to be as short as possible (4mm in this test) so as not to affect the waveform of the measured strain. A PCB instrumented hammer with a plexiglass plastic tip was used to impact the long bar. The same Measurement Group model P-3500 bridge amplifier was used in quarter bridge configuration to measure the strain. The wire connecting the strain gauge to the bridge was of twisted pair construction 1.0 metre long. The output signals from the hammer and bridge were analyzed on either Norland 3001 or B & K model 2032 dual channel analyzer. Figure 6.4 show a sketch of the test set up.

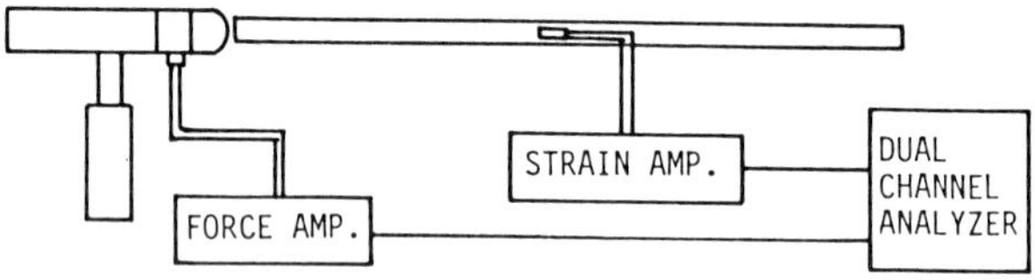

Figure 6.4: Test set up for long bar calibration

Figure 6.5 shows time histories of both the input force function f(t) and the strain measured at distance 1.29m from the impacted end of the bar (e(1.29,t)).

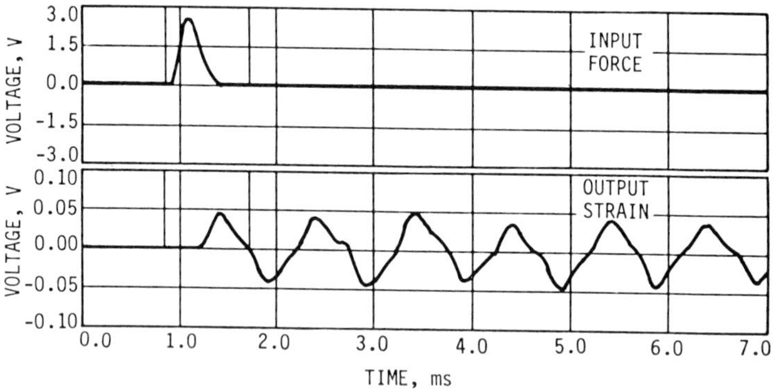

Figure 6.5: Input impact force and output strain time histories

If we have an ideal measuring system, the frequency response function taking the first half cycle of the strain as output and the force as input should be one of constant amplitude. The phase shift should be linear lag with slope 360 T1 degrees per hertz, where T1 is the time delay between the anvil and the strain gauge.

The elastic wave propagation speed of the used aluminium bar, equation (6.10), is calculated as follows:

$$c = \sqrt{E/\rho} = 5138 \text{ m/s}$$

and consequently delay T1 is

T1 = 1.29/c = 251 microseconds

For the purpose of calibrating the measuring system, the frequency response function between the output strain and the input impact force at the same point should be considered. The same FRF can be obtained by calculating the FRF between the input force and the strain at distance x with the later function shifted in time by x/T1 seconds. Another point to consider is that we normally collect a sample which includes the strain history while the wave is travelling in both directions of the bar. For this reason, only the first part of the strain history need to be used in calculating the FRF.

Figure 6.5 shows the imput (upper part) force function and the output (lower Part) strain function. A rectangular window on both graphs marks the part used in calculating the FRF. Figure 6.6 shows the FRF of the measuring system obtained by the method outlined.

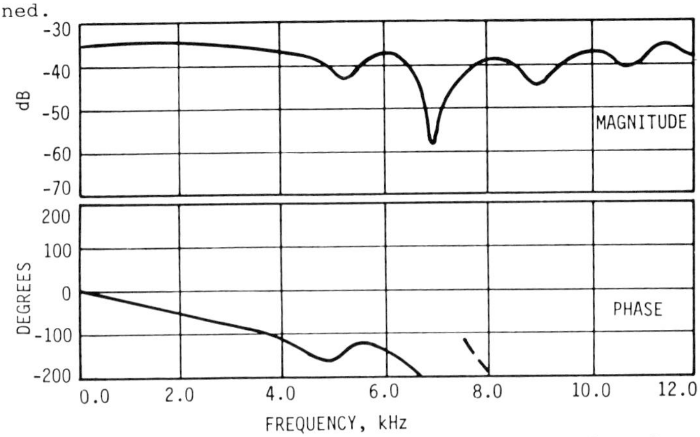

Figure 6.6: FRF of the measurement system using long bar calibration

The measured FRF shows a drop in the magnitude ratio (upper part) of nearly 3 dBs in the range of 0 to 4000 Hz. In the same region, the phase part shows a linear phase lag at a rate of 0.028 degrees/Hz. Beyond this range phase shift starts to deviate from being linear. The magnitude ratio exhibits irregularity at 6912 Hz. At the same frequency the phase shift is not defined altogether. This is not the actual FRF of the measuring system, rather, it is the effect of having the input and output spectra having low values at this frequency. Assuming the impact function to have a half sine profile and duration of T seconds, the corresponding spectrum has zero values at 1.5/T, 2.5/T, 3.5/T, etc. Hz. In our case, a 500 microsecond duration implies zero level power spectra at 3000 Hz, 5000 Hz, 7000 Hz, etc. Actually, for the impact profile tested it appears that this problem shows forcefully at the 6912 Hz. So, it is advised to use the FRF in the 0 to 4000 Hz range exclusively. If a wider range is to be defined, a shorter pulse should be used.

Concluding, we may say that the calibration of the strain gauge measuring system is described by FRF shown in Figure 6.1(b). This FRF can be used to calibrate modal analysis data that uses this specific combination of hammer and bridge amplifier.

CONCLUSIONS

Calibration of a strain measurement system for modal analysis requires the determination of the frequency response function of the measuring devices. A simple method of using an electronic circuit to simulate a rectangular pulse of strain variation was used to obtain an accurate estimate of the FRF of a strain measuring system. The use of free-free beam calibration suggested earlier in literature was shown to be inappropriate. An alternative calibration scheme that uses the strain measured at some point of a long bar impacted at its end was used for strain measurement system calibration. This later method has the advantage of a predetermined dynamic response of the bar, so compensating for the time delay in the bar itself is possible.

ACKNOWLEDGEMENTS

This research was conducted in the vibration laboratory facilities of the Department of Engineering Science and Mechanics and the Engineering Research Institute of Iowa State University. Their support is appreciated.

REFERENCES

[1] Okudo, N., S. Tanabe and T. Hirano, "Basic Consideration in the Measurement of Dynamic Strain Mode Analysis", Proc. Spring 1986 Mtg., New Orleans, LA, 245-249, June 1986

[2] McConnell, K.G., "A Correction of A Strain Gauge Calibration Procedure for Modal Analysis", SEM Fall Mtg., Colorado, CO, November 1986

[3] Seireg, A., "Mechanical Systems Analysis", International Textbook Company, 1969

7 Investigation of masonry arch bridges using acoustic emission

R.Royles & A.W.Hendry

SUMMARY

A description is given of the use of acoustic emission (AE) monitoring in the full scale static load testing to failure of two stone masonry arch bridges. The first was an approximately parabolic arch and the second semi-circular. The instrumentation for each bridge and the general test procedure is outlined. The general load-deformation behaviour of the bridges is discussed and the acoustic activity emanating from the bridges under load testing is considered in relation to the overall behaviour. Implications for the use of acoustic response in the integrity testing and assessment of masonry arch bridges are presented.

Key words : Acoustic emission, masonry arch bridges

INTRODUCTION

In the UK there is a large number of masonry arch bridges in brickwork and stone within the canal, rail and road networks. The structural arrangements cover a wide variety of types from small single span culverts to large multi-span bridges, most of which were constructed in the eighteenth and nineteenth centuries. In the main, the performance of these bridges has been excellent especially when it is taken into account that there have been marked increases in traffic densities and loadings for road bridges in particular. This has given rise to a need for reliable methods of integrity and strength assessment of such bridges.

Various strength prediction methods have been proposed for masonry arch bridges and some associated full scale and model load tests have been carried out. However, the lack of agreement between experiment and theory has cast doubt over the accuracy of current analytical methods. Recently a programme of approximately ten loading tests to failure was commissioned by the Transport and Road Research Laboratory (Department of Transport) and supported by the Scottish Development Department to build up information about the behaviour of masonry arch bridges on full scale. The two tests described have formed part of that programme.

In addition to examining the actual response of masonry arches to static line loading and comparing it with theoretical predictions [1,2] a further main objective of the two tests was to gather information about the development of cracking, displacements and associated acoustic emission (AE). It is the latter aspects of the investigation which are highlighted below.

BRIDGE DETAILS

The two bridges chosen for testing were in Ayrshire. The first was crossing the Girvan water at Bridgemill on the A77, Ayr to Stranraer trunk road, and the second at Bargower on the A76, Kilmarnock to Dumfries trunk road, crossing the Cessnock Water south of Kilmarnock. Both bridges became available for testing as a result of road realignment schemes rendering them redundant and were of a similar age - Bridgemill built in 1869 and Bargower in 1859 using red sandstone. Results from material control tests on the stone are given in Table 7.1.

Table 7.1

Characteristics of the red sandstone

Bridge	Mass Density kg/m^3	Crushing Strength MN/m^2	Compressive Young's Modulus MN/m^2
Bridgemill	2100	43.8	15000
Bargower	2730	33.3	14100

Geometrical information for the two bridges is summarised in Table 7.2.

Table 7.2

Geometrical data

Bridge	Span at Springings m	Rise at Crown m	Span to Rise Ratio	Overall Width on Vault Centre Line m	Arch Shape	Skew degree of arc	Voussoirs No.	Voussoirs Depth mm
Bridgemill	18.30	2.85	6.4	8.30	Parabolic	0	62	710
Bargower	10.00	5.00	2.0	9.25 (parapet) 8.68 (vault)	semi-circular	16	53	558

The leading dimensions for Bridgemill are given in Figure 7.1 and for Bargower in Figure 7.2.

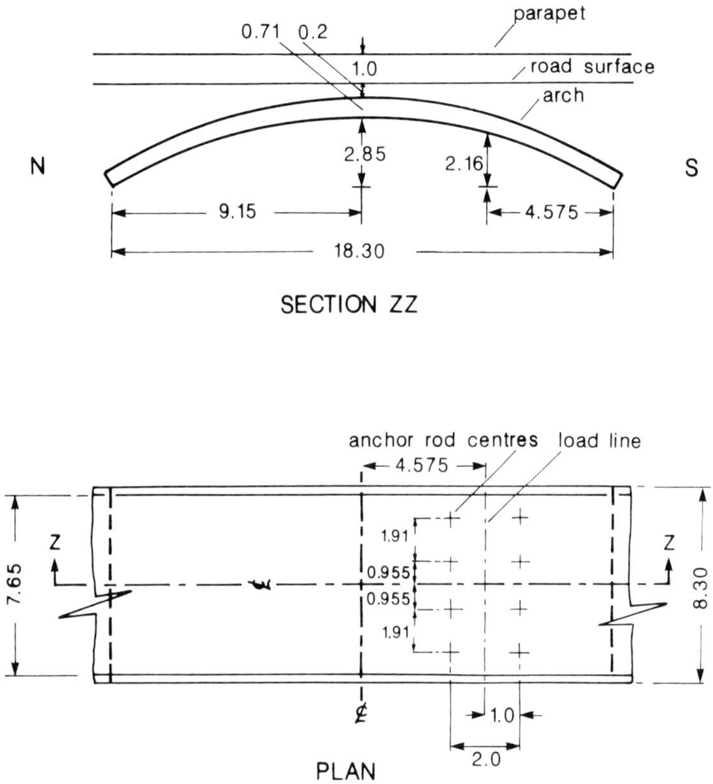

Figure 7.1: Bridgemill dimensions (m)

At Bridgemill the arch was approximately parabolic with the voussoirs cut to a regular shape and with zero skew. The whole condition of the bridge was good with the spandrel walls and arch vault being free from significant cracking and only one of the parapet walls being out of alignment - displaced outwards by a maximum of 25mm.

The vault of the Bargower arch was built in accurately cut stone blocks angled to accommodate the 16° skew. The springings of the arch were some 6m above river level and rested on coursed masonry retaining walls. Some haunching was detected by coring through the vault near the springings and extended to approximately 1.5m above the springings. Spalling of individual stones in the vault, damage to the stonework in part of the abutment wall, outward rotation of spandrel and parapet walls, and a significant longitudinal crack in the vault approximately 1.25m in from the west skewed face of the bridge and extending across almost the whole span were all in evidence. This led to the conclusion that its pre-test condition was only moderate.

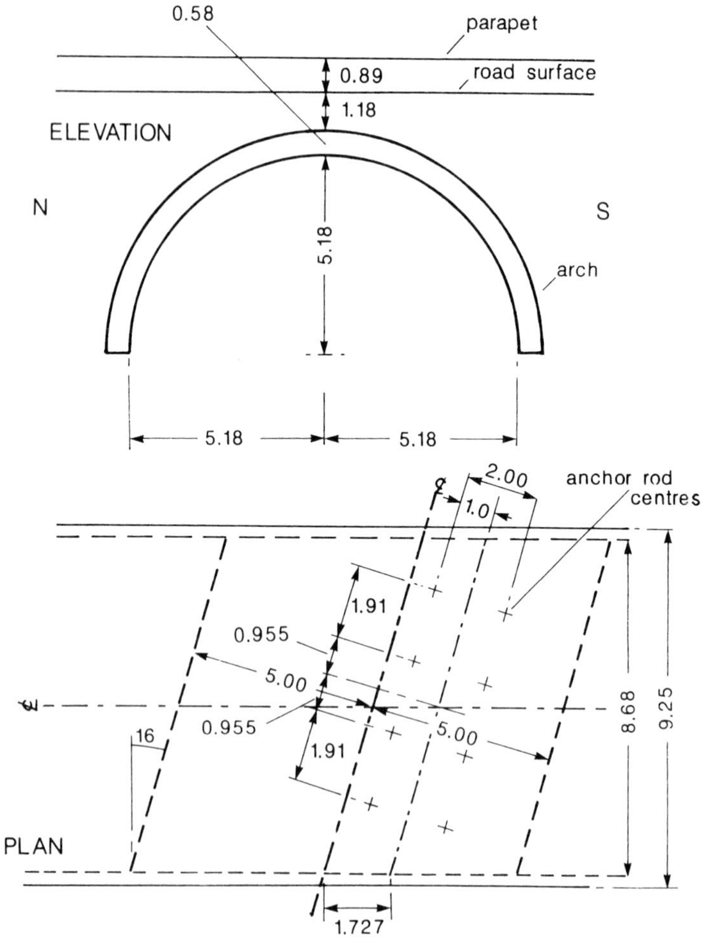

Figure 7.2: Bargower dimensions (m)

ARRANGEMENTS FOR TESTING

The carriageways of both bridges had a North-South alignment. Transverse vertical line loading was provided in each case using a system of four hydraulic jacks reacting against cross beams above the bridge deck. A cross beam was pinned by two 50mm diameter steel rock anchors passing through pre-drilled holes in the deck and secured in the rock bed below the river. A 2000kN load cell was placed between each jack and cross beam. The safe working load of each rock anchor was 750kN, permitting the application of loads up to 6000kN.

Preliminary analyses identified quarter span loading as being the most arduous position for Bridgemill contrasting with the one third span position for Bargower. In the latter case the line loading followed the skew and was parallel to the vault centre line.

Horizontal and vertical displacements were measured by LVDT transducers and a precise triangulation theodolite method at Bridgemill for the springings, crown and quarter span points on both east and west elevations of the arch. At Bargower only the surveying technique was employed using targets at the springings, crown, unloaded quarter point and the load point (third span point) on both arch elevations. LVDT transducers were used at Bargower to monitor movements at six points along the longitudinal crack in the vault.

Acoustic emission transducers were disposed in four positions around the arch rib in the west elevation of each bridge at estimated hinge locations. At Bridgemill four pairs of sensors were employed - one near the top and the other near the bottom of a voissoir, each sensor being coupled via a stiff grease to the squared end of a steel guide embedded in the stone. Only four sensors were used at Bargower, one in the middle of the voussoir adjacent to a predicted hinge position. In this case sensors were coupled direct on to the smoothed stone face via stiff grease. The sensors were Dunegan type S9204 each with a preamplifer type 1801 set to give a gain of 40dB. The quality of the coupling between a sensor and the structure was checked using a pencil splitting technique as was the order of signal attenuation in stone masonry. The background noise was nowhere in excess of 13dB at Bridgemill and 19dB at Bargower relative to 1 μV at a sensor. Filtering at the level of 30dB was set for each sensor to avoid recording lower intensity signals. Rise times to a signal peak were computable from a 20dB threshold level.

The disposition of AE sensors for each bridge is shown in Figure 7.3.

TEST PROCEDURE

(a) Bridgemill (May 1984): After establishing datum readings for all transducers loading increments of 200kN were applied

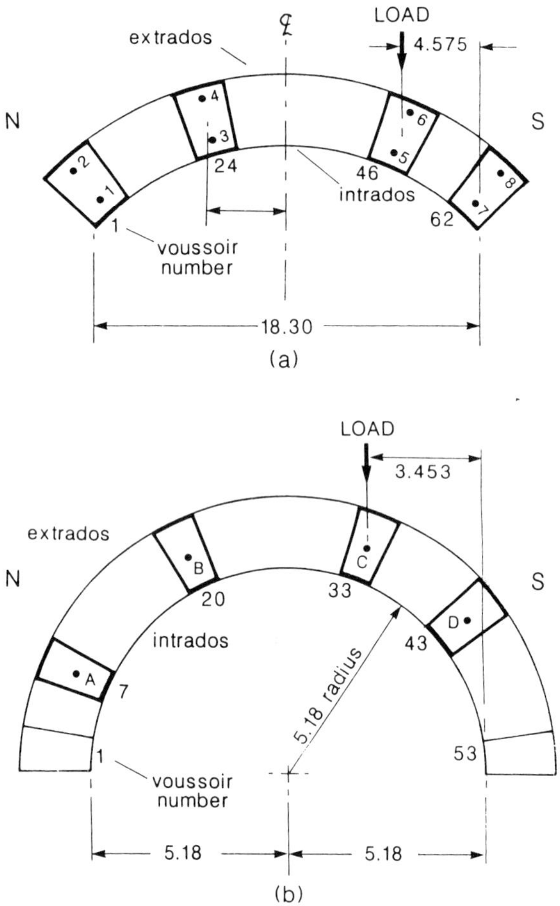

Figure 7.3: Location of AE sensors (a) Bridgemill (b) Bargower • sensor (all dimensions in m)

up to 2400kN at which level jack stroke was exhausted and unloading took place to allow the jacks to be adjusted for the excess deflection. Re-loading was resumed continuously up to 1800kN and then in increments of 200kN to 3000kN at which stage unloading was necessary again to reset the jacks owing to stroke limitations. On resumption of loading a maximum of 3100kN was achieved and despite six further loading cycles this maximum was never exceeded with the deformations becoming progressively large.

At each load increment the load was held for a few minutes whilst the theodolite measurements were taken. Displacement transducers were monitored instantaneously with a Solatron Orion data logging system at the beginning and end of a load holding period. The AE sensors were monitored continuously by a Dunegan

8000 computer system through the first two loading cycles.

(b) Bargower (June 1985): A similar procedure was adopted to that followed at Bridgemill a year earlier and 200kN load increments were applied in the first cycle up to 4600kN before unloading to adjust jack positions to give more stroke. Displacements of the bridge arch were measured on both elevations by the precise surveying method and displacement transducers were located to monitor changes in the longitudinal crack. A fault developed in the Dunegan 8000 system after the first cycle and no AE data were obtained in subsequent cycles. A further five load/unload cycles were carried out and in the fourth one of these an overall maximum load of 5600kN was achieved. Thereafter the peak loads in subsequent cycles decreased.

LOAD DEFORMATION BEHAVIOUR

The deflection responses of the bridges to vertical line loading are represented typically in Figure 7.4a and b for Bridgemill and Bargower respectively by the mean vertical displacement at the load point. It can be seen that the smaller span Bargower bridge was the stiffer of the two, and both behaved non-linearly becoming progressively less stiff with increasing load.

In general at Bridgemill with the southern quarter point loaded, the abutments moved out horizontally and rose to the north and settled to the south. The quarter points moved northwards horizontally and the loaded one moved down whilst the unloaded one rose. The crown tended to move northwards horizontally and rise.

At Bargower the abutments moved negligibly, the load point only moved downwards, the crown moved northwards and downwards and the unloaded quarter point moved northwards but did not displace vertically. In both bridges the deformations were small even at maximum load, being much less than 1/3 x depth of a voussoir. Total collapse was not induced in each test.

ACOUSTIC RESPONSE

Acoustic emission was considered in terms of events, each event comprising a number of counts [3]. When the signal from a transducer exceeded a pre-set threshold level (in these tests 30dB relative to 1 μV at the transducer is 31.7 μV) an event was deemed to have been initiated, and to have finished when the signal failed to exceed the threshold within a pre-set time interval (200μs in these tests) from the most recent excursion above that level. The number of excursions above the threshold in a rising manner within an event were defined as counts. Event size was described by peak signal amplitude and the number of counts within it.

Attempts were made by means of pairing signals from sensors to locate crack sources but triggering of more than one sensor

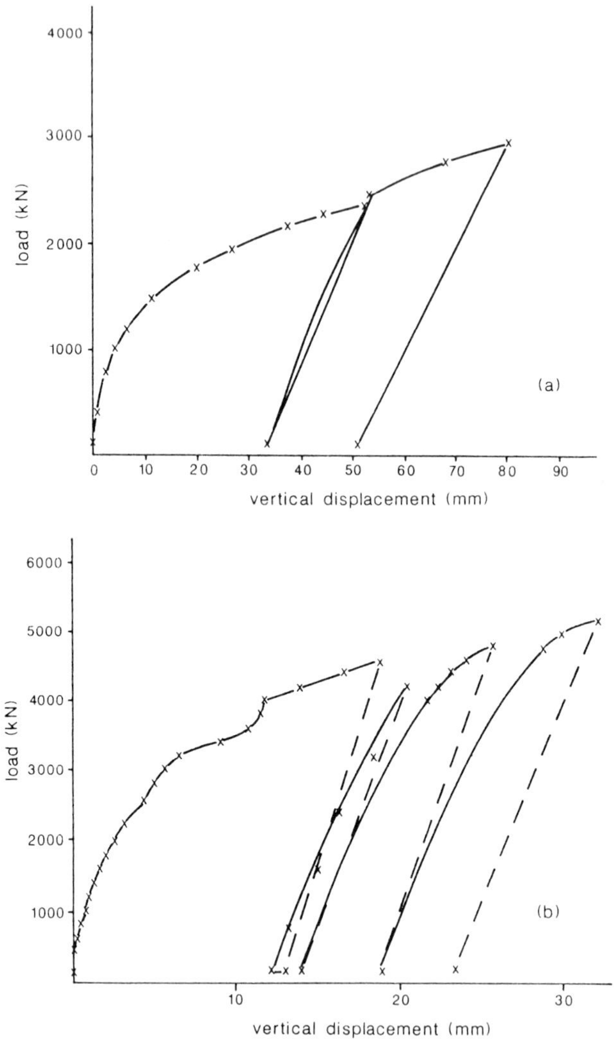

Figure 7.4: *Mean vertical downward displacement at load point (a) Bridgemill (b) Bargower*

by the same signal occurred only in a few instances. The location technique was similar to that employed in seismic studies. All data were screened so that any signals due to reflection were rejected.

(a) Bridgemill: The emission grew at an increasing rate at

all sensors as the load increased with the major part of the activity emanating from the load point region and north of the crown (Figure 7.3a). The general pattern of emission was such that more tended to come from zones associated with compression rather than tension and was consistent with the observed hinging pattern near sensors 2 and 6 on the extrados and 3 and 7 on the intrados.

On a cumulative basis for all sensors the development of emission during the first two loading cycles is shown in Figure 7.5a. Significant emission started at 1000kN. The unloading from 2400kN in the first cycle was followed by silence until the 1500kN load level was attained during re-loading in the second cycle, with the load-emission curve rejoining that of the envelope formed by extrapolation of the load increasing curve of the first cycle.

The ratio of the load level at re-emission to that of the previous peak is termed the "felicity" ratio. A high felicity ratio of the order of unity is indicative of little permanent damage having occurred whereas the opposite is indicated by a felicity ratio much less than unity [4].

The collective felicity ratio indicated by Figure 7.5a is approximately 0.63 and is associated with some permanent deformation of the bridge (Figure 7.4a). On the basis of the felicity ratio, results from individual sensors suggested that most damage had occurred at the load point up to the end of the first cycle, which might be expected.

The time of rise to the initial peak amplitude in an event is shown in Figure 7.6a for all events and can be seen to have been predominantly less than 100 µs which suggests that activity was due to tensile cracking rather than frictional rubbing. The latter produces elastic waves of very much lower frequency than those associated with tensile cracking.

(b) Bargower: The voussoir size was much less than that at Bridgemill and consequently favoured the use of one sensor at a predicted hinge location on the arch. The sensor positions near the abutments (Figure 7.3b) were influenced by the presence of stiff internal haunching [2].

Again, as load increased the emission from all sensors grew at an increasing rate. The activity came predominantly from the load point region and the sensor nearest to the south abutment (Figure 7.3b).

Unfortunately due to an instrument malfunction it was not possible to monitor acoustic activity beyond the first load cycle.

The accumulated emission from all sensors is shown relative to the load for the first cycle in Figure 7.5b and the rise times for the peak amplitudes of all the events are shown in Figure 7.6b. This latter distribution suggests that, as at Bridgemill, the events were of the high frequency type associated with tensile cracking. Emission occurred significantly above the 1600kN load level, becoming distinctly non-linear with load above 2000kN. A surge in activity occurred at 3400kN which corresponded with sudden movement in the bridge at that load level (see Figure 7.5b). The loading rate with time was fairly uniform up to the

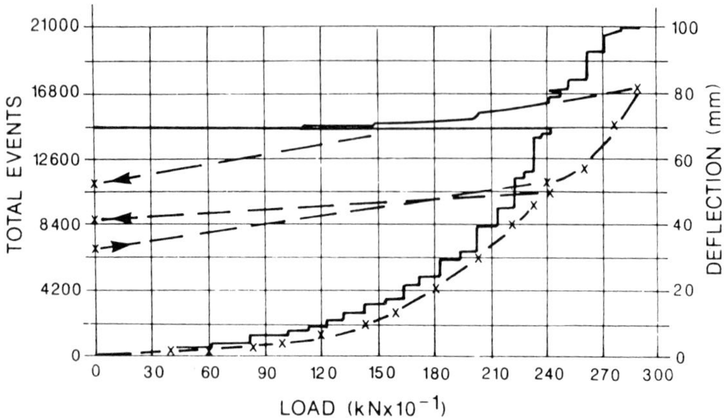

(a)

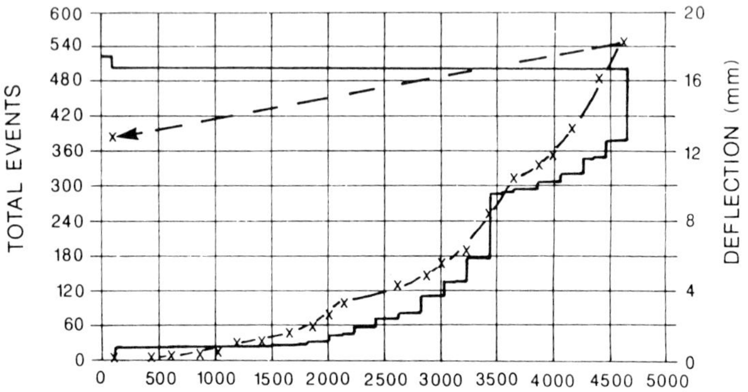

(b)

Figure 7.5: *Deformation responses (a) Bridgemill (b) Bargower*
—— *cumulative AE from all sensors*
X *mean vertical downward deflection at load point*

maximum load of 4600kN in the first cycle but the corresponding emission rate was non-linear above 2000kN becoming temporarily very rapid with the bridge movement at 3400kN and at the first cycle peak of 4600kN.

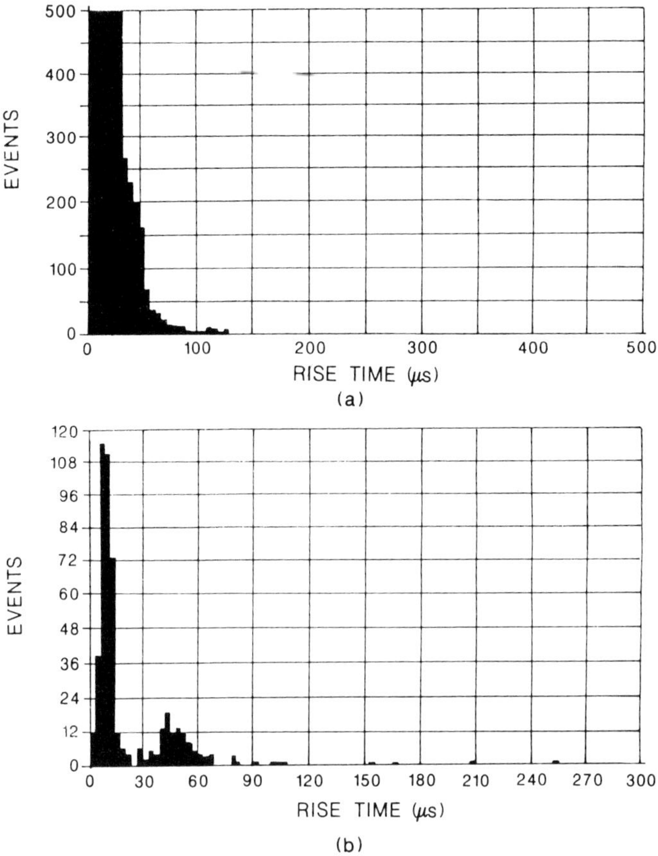

Figure 7.6: *Peak amplitude rise time distribution (a) Bridgemill (b) Bargower*

DISCUSSION

There was a marked contrast in size between the acoustic signatures of the two bridges. Bridgemill produced 21000 events in two cycles as it was taken to 97% of its ultimate load capacity, and 14700 events in the first cycle to a peak of 81% of its ultimate load. By comparison Bargower produced only 530 events in the first cycle to a peak at 82% of its ultimate load.

The quieter nature of the Bargower bridge coincided with its stiffer load-deformation characteristics (Figure 7.4a and b). However, it might have been partly a consequence of the poorer pre-loading condition of that bridge. It could be expected that a bridge in good condition and of the same sort of material as one in poor condition would emit more during a load test to failure.

The plots of load and deflection shown in Figure 7.5a and b demonstrate for both bridges the similarity between that characteristic and the load - total events relationship. This suggests quite good correlation between AE and deflection, and that AE is a parameter which could be used to represent deformation behaviour in masonry arch bridges.

It is quite interesting to note that the onset of non-linear behaviour with load in terms of AE occurred at approximately 1/3 x ultimate load for both bridges.

Subsequent re-loading after the first load cycle, in the case of Bridgemill, did not produce any renewed emission until a load level a little in excess of 1/3 x ultimate had been achieved, indicating that no further permanent damage had occurred in that range. Although hinge mechanisms did not develop fully in the two tests [1,2] activity from the individual sensors did reflect the tendency to form hinges in those localities.

IMPLICATIONS FOR INTEGRITY TESTING

The features of the AE response in which non-linear behaviour with load was tending to occur at 1/3 x ultimate load capacity would suggest that this might form the basis of a non-destructive form of load testing for masonry arch bridges to determine their load carrying capacity.

The sensors are relatively small and easy to mount and the instrumentation is portable. In combination with a loading vehicle the equipment could be employed to determine the onset of non-linearity in the acoustic signature of a bridge. The corresponding load level would represent a value having a factor of safety relative to ultimate carrying capacity. The tests described above indicate a factor of safety of the order of 3 would pertain and that this could have been adopted as a sensible safe working load. However, more test data are required to confirm this value for an acceptable factor of safety.

It would seem preferable to seek the overall cumulative AE characteristic of a bridge and to use a minimum of eight sensors, four on each elevation of an arch, in order to establish the non-linear stage relative to load. The disposition of sensors on an arch would be guided by some suitable prediction method for hinge location taking into account the shape and initial condition of the bridge [5,6]. Once emission has occurred it will not recur normally unless there is some form of self-healing taking place in the material. Consequently AE is not capable of directly accounting for previous load history but could be used to predict the reserve of strength in a masonry arch bridge. This non-linear point on an AE - load characteristic would be synonymous with the onset of non-linearity in the energy release rate of an elastic body, which is a feature of linear elastic fracture mechanics [7].

CONCLUSIONS

1. Acoustic emission is a parameter that could be used to represent deformation behaviour in masonry arch bridges.
2. Hinge formation in masonry arches is reflected in activity from AE sensors.

ACKNOWLEDGEMENTS

This work was commissioned by the Department of Transport - Transport and Road Research Laboratory and the Scottish Development Department through Strathclyde Regional Council, and to all these agencies the authors offer their thanks for the support and encouragement in publishing this work. Sincere gratitude is expressed also to the secretarial and technical staff of the University of Edinburgh, Department of Civil Engineering and Building Science for all their assistance.

REFERENCES

[1] Hendry, A.W., Davies, S.R., and Royles,"Test on a stone masonry arch at Bridgemill - Girvan", CERTI Report, University of Edinburgh (Sept. 1984) and Department of Transport, TRRL, Crowthorne Contractor Report 7 (1985)

[2] Hendry, A.W., Davies, S.R., Royles, R., and Ponniah, D.A., "Load test to collapse on a masonry arch bridge at Bargower, Strathclyde", CERTI REPORT, University of Edinburgh (Sept. 1985) and Department of Transport, TRRL, Crowthorne, Contractor Report 26 (1986)

[3] Hendry, A.W., and Royles, R., "Acoustic emission observations on a stone masonry bridge loaded to failure", Structural Faults 2, 30 April-2 May 1985, Institution of Civil Engineers London, Engineering Technics Press, Edinburgh (May 1985), pp. 285-291

[4] Fowler, T.J. and Gray, E., "Development of an acoustic emission test for FRP equipment", ASCE Convention, Boston, Mass. 2-6 April 1979

[5] Davies, S.R., "Microcomputer analysis of the influence of certain parameters on the assessment of masonry arch bridges", Structural Faults 3, 7-9 July 1987, Goldsmith's College, London, Engineering Technics Press, Edinburgh (July 1987) pp. 305-309

[6] Falconer, R. and Jones, K., "The assessment of masonry arch bridges using commercial software", Structural Faults 3, 7-9 July 1987, Goldsmith's College, London, Engineering Technics Press, Edinburgh (July 1987), pp. 337-341

[7] Parker, A.P., "The mechanics of fracture and fatigue - an introduction", E. & F.N. Spon, London (1981)

8 Into the 1990's with the acoustic strain gauge

D.W.Begg & J.E.Butler

SUMMARY

Current developments in acoustic or vibrating wire gauge construction and its associated instrumentation have brought about renewed interest in this method of strain measurement. It is now possible for a single acoustic system to monitor both dynamic and long term static effects, this can lead to a large reduction in instrumentation costs for projects where electrical resistance strain gauges are used to monitor dynamic effects and acoustic gauges are used to monitor very long term effects.

INTRODUCTION

The discovery of the principle upon which the acoustic gauge is based was first published by Martin Mersenne in his 'Harmonicorum Liber' in Paris in 1636. The use of a vibrating wire for the measurement of strain did not appear until almost 300 years after Mersenne's work with vibrating strings, when in 1928 a paper by Dawidenkow [1] was published. Early gauges were used for measuring surface strains, although Dawidenkow suggested their suitability for embedment in concrete.

THE VIBRATING WIRE PRINCIPLE

<u>Marsenne's Law</u>

'The frequency of the fundamental mode of transverse vibration of a string varies as the square root of the tension, inversely as the length and inversely as the square root of the density'.
Referring to Figure 8.1a

$$f = \frac{1}{2L}\sqrt{\frac{T}{M}} \quad \text{Hz}$$

where
f = fundamental frequency of transverse vibration

T = tension in the string
L = length of the string
M = mass per unit length of the string

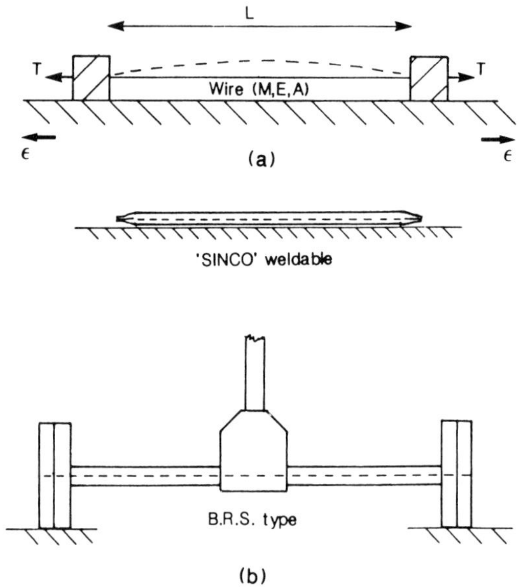

Figure 8.1: *(a) The Vibrating Wire Principle*
(b) Gauge configuration examples

In an acoustic gauge the vibrating string is in fact a steel wire which is assumed to behave in the same manner as a string; that is its resistance to bending is considered to be negligible.

The resulting change in frequency of oscillation of a wire caused by it undergoing a change in strain is thus obtained by applying Hooke's Law.

$$\text{Change in Strain} = k.(f_0^2 - f_1^2)$$

where
K = constant relating to wire properties
A = cross sectional area of string
E = Young's modulus of the string

$$K = \frac{4L^2M}{AE}$$

It can be seen from the above that the device can be assumed linear in terms of frequency squared. It is this parabolic as opposed to linear relationship that creates difficulties in the acceptance of the device by some.

STATIC MEASUREMENTS AND TRANSDUCERS

The most significant advantage vibrating wire transducers have over analogue measurement methods is its high tolerance to earth leakage and long cable lengths. Reductions in signal level are easily recovered with suitable amplification. The signal is also ideally suited for telemetry.

Typical gauge configurations can easily be adapted to the requirements of the specific monitoring task. Some examples are shown in Figure 8.1b.

Static instrumentation

Early instrumentation was unable to measure frequency directly, strain measurements were made by direct comparison of signals from the active gauge with a control gauge [2] using 'beats' and later Lissajous figures [3]. The availability of integrated circuits brought about a large reduction in size of recording devices together with the ability to measure frequency (or time period) directly.

Plucked systems

These systems have the advantage that they only require one coil at the gauge. The recorder initiates a DC electric pulse through the coil to 'pluck' the wire, then after a suitable delay of several microseconds either frequency is measured over one second or the time for a number of cycles is measured. The inclusion of a microprocessor allows the reading to be displayed directly in engineering units, e.g., microstrain, N/mm^2,m. Temperature compensation is now also possible where acoustic gauges are attached to materials having a different thermal expansion coefficient to that of the wire.

Transducer applications

The advantage of the vibrating wire principle make it particularly suited to long term and remote measurements. For this reason a recent rock anchor scheme in Kuwait specified their use in the annular ring load cells used to monitor these anchors over the next 20 years. The cells used for this project (Figure 8.2) incorporate three vibrating wires and are completely sealed from the environment using a welded cover as opposed to the more normal 'O' rings. The instrumentation developed for monitoring these cells is capable of storing the individual callibration data for each of the three wires, in the many cells, in Non Volatile

Ram thus allowing the cells to operate if one or even two of the wires fail during the life of the rock anchors. An additional feature with this instrumentation is its ability to compensate for both thermal coefficient mismatch and material modulus variation with temperature.

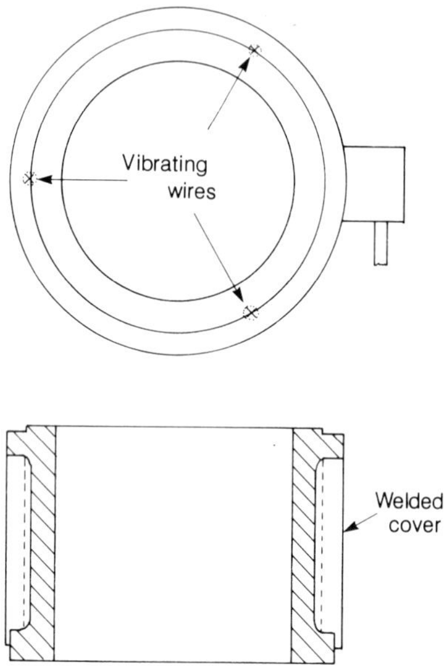

Figure 8.2: Acoustic annular ring load cell

Dynamic measurements

Early attempts at making dynamic measurements with an acoustic gauge used a single coil arrangement [4,5]. The amplitude of wire oscillation was monitored and when this decayed to a predetermined level the wire was plucked again.

The most economical and simple method of monitoring dynamic effects using the vibrating wire principle makes use of a positive feedback amplifier together with both excitation and pick-up coils to maintain the wire in continuous oscillation. The frequency signal thus generated may be monitored by any counter/timer instrument. The main disadvantages with using normal frequency or time period measurement instrumentation for monitoring is that it is extremely slow and the data obtained has to be reprocessed into meaningful results, e.g., microstrain. Interfacing of plucked systems has allowed a maximum of less than 10 readings per second. This limit restricts its use in dynamic applications.

Data logger interfacing

For the acoustic gauge to become more widely used in the field of structural monitoring it is necessary for its output to be easily interfaced with standard data logging equipment. As a Solartron Orion data logger was already available within the department the ability to use this for acoustic gauge monitoring has been investigated.

Attempts to use the Orion in conjunction with a plucked vibrating wire system failed due to the interference of the pluck pulse with the frequency measurement. Continuous wire vibration using a simple audio amplifier allowed the use of the Orion's optional frequency module, whilst this is a very simple method it proved to be exceptionally slow as the Orion required 1 second time delay between successive readings.

A high resolution frequency to voltage converter has been designed which enables the Orion to log a DC voltage directly proportional to frequency. Calibration of this device has shown its exceptional linearity (see Figure 8.3b). This enables the Orion to log frequencies without the optional frequency module at much faster rates.

The programmable conversion facilities, Scalar or Polynomial, available to the Orion user allow processing of this voltage so that a frequency or strain may be displayed directly or recorded on magnetic tape. In this way the Orion is able to log dynamic or static strains from an acoustic gauge. Communication between the Orion and a micro-computer is possible should it be required [6].

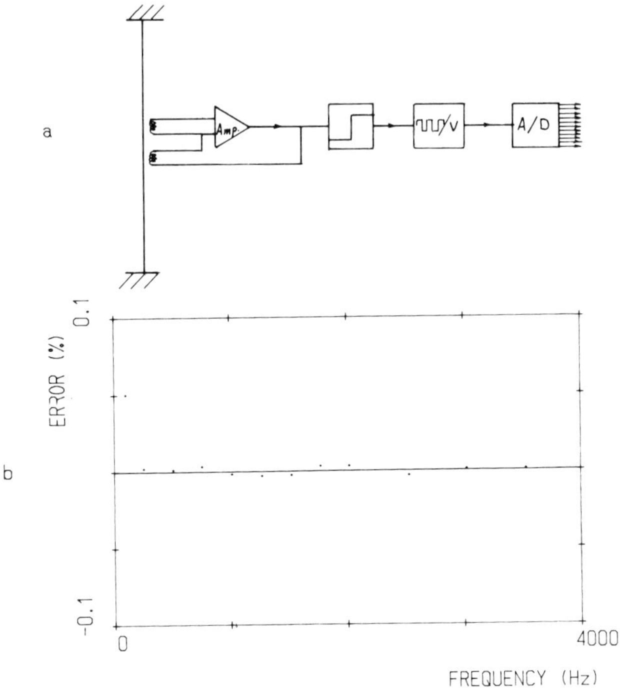

Figure 8.3: *(a) Example of continuous excitation interface*
(b) Frequency to voltage linearity

Computer interfacing

The ability of microcomputers to process signals from acoustic gauges has been investigated and it has been shown that they can provide an effective means of logging strains. Two computers have been used with different methods of signal processing.

1. BBC Micro

In this case a voltage proportional to frequency was input to the computer's analogue port. This A to D conversion was calibrated using a signal generator with timing of 1000 cycles for checking the actual frequency. The resolution of frequency measurement was limited by the conversion being 12 bit, and as the output from the frequency to voltage converter was approximately one volt for 1000Hz, the maximum voltagc allowed by this port restricts the frequency range to 1800Hz. This limit would be acceptable for many vibrating wire applications.

Programs have been written in Basic for display of frequency or strain and also for logging data and subsequent post-processing. The integration period of the F to V converter restricted the speed at which readings could be taken to 20 per second although the computer could scan 4 channels at this rate.

2. RM Nimbus

In an attempt to increase logging speed and improve frequency resolution an interface has been designed which utilises a 16 bit Binary counter and 10 MHz crystal oscillator, under direct computer control, to measure the time period. A machine code program is able to both display frequency and log readings at speeds of up to 500 per second. This rate of sampling makes it possible to monitor any naturally occuring dynamic effects on real structures.

Results of dynamic tests

Figure 8.4 shows the results of tests on a stainless steel 12.5mm square bar specimen loaded in a dynamic test machine from 10kN compression to 10kN tension with a forcing frequency of 0.25Hz. The data was logged using the Orion, utilising its facility to store data directly on magnetic tape. The example of triangular waveform is included to highlight any nonlinearity in going from tension to compression with rigid gauge construction.

Figure 8.5 shows results from similar tests logged by a BBC micro. These results indicate the speed at which data can be logged using an inexpensive microcomputer. A small amount of noise is present on this data, which was considered acceptable in this case as the strain range is quite low.

Figure 8.6 is an example of results from the NIMBUS micro showing the improved frequency response. It is possible using this interface to log readings at up to 500 per second. However, the simultaneous display of results slows this down to approximately 200.

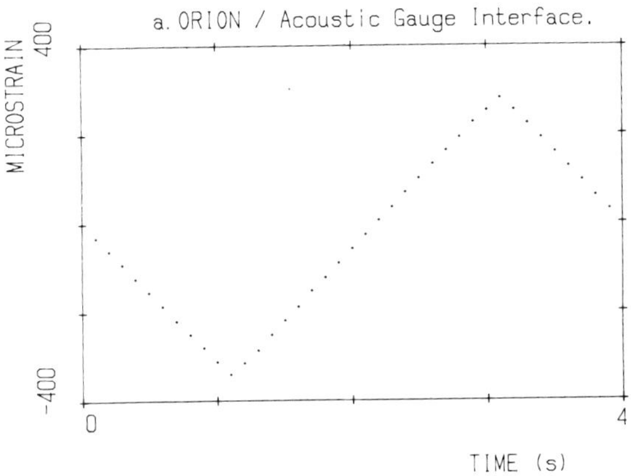

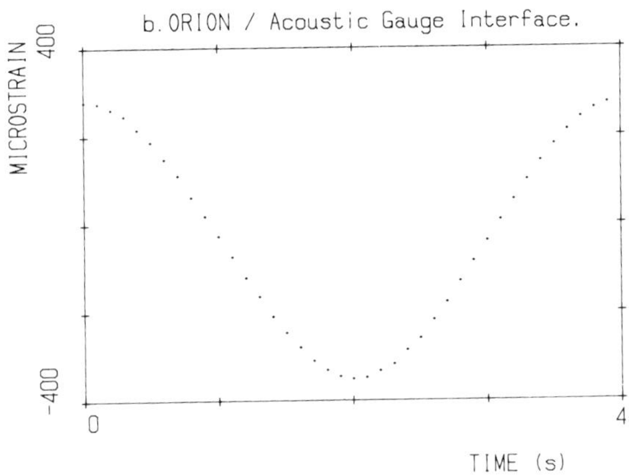

Figure 8.4: *(a) Orion/acoustic gauge interface*
(b) Orion/acoustic gauge interface

DISCUSSION

The results from these tests show the ability of the acoustic gauge to monitor dynamic effects in structures. The availability of the microprocessor for inclusion in, or interfacing with, vibrating wire instrumentation dispenses with the need to convert a displayed frequency into engineering units.

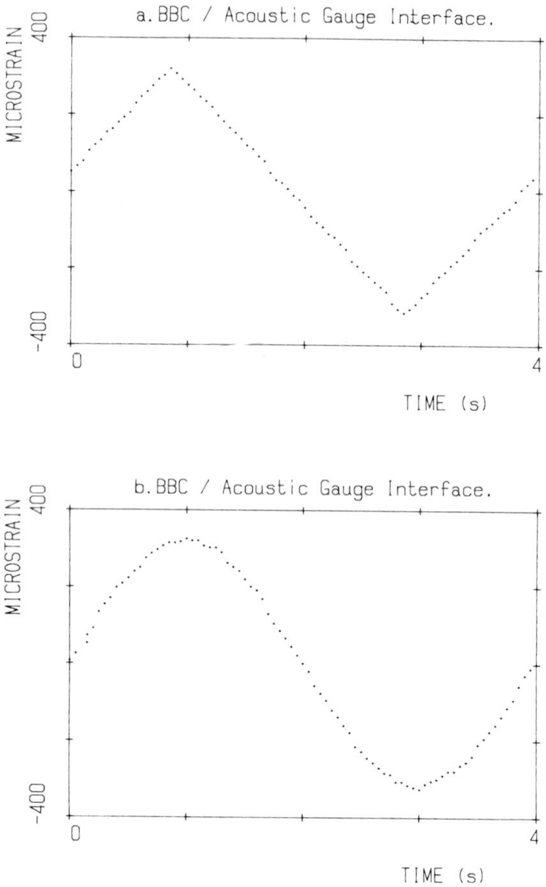

Figure 8.5: (a) BBC/acoustic gauge interface
(b) BBC/acoustic gauge interface

CONCLUSION

The use of the vibrating wire principle for the measurement of strain could receive a new lease of life from the application of the principles outlined above. Monitoring tasks requiring embedment gauges capable of both dynamic and static measurements are possible with the acoustic gauge if some thought is given to gauge configuration; however it is essential that the instrumentation developed for these systems is compatible with standard data loggers.

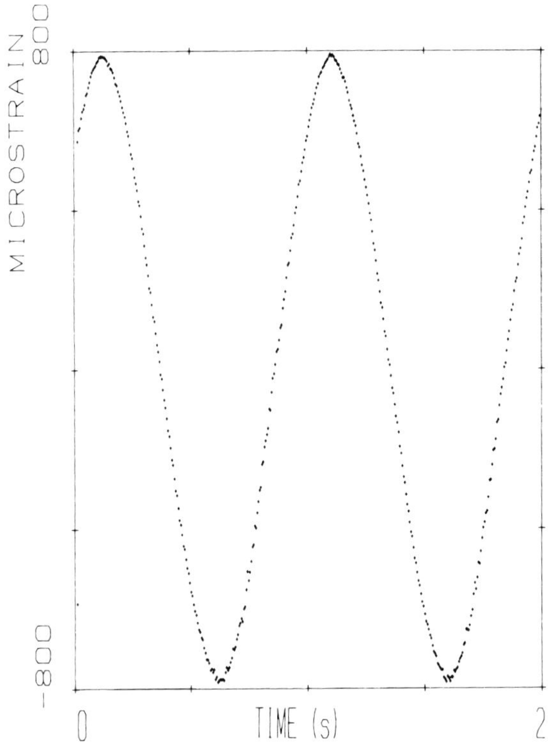

Figure 8.6: NIMBUS/acoustic gauge interface

ACKNOWLEDGEMENTS

Appreciation is hereby recorded for the helpful cooperation extended by Strainstall, a division of Crystalate Electronics Ltd. for their continued support for work in this area, and for the valuable advice and assistance provided by the technical staff of the Department and 3D (Digital Design and Development Ltd).

REFERENCES

[1] Dawidenkow, N., "An Acoustical Method of Measuring Stress", Journal of Applied Physics, Leningrad, 1928, 5, pp. 37-46
[2] Davey, N., "Measuring the Existing Stresses in Masonry Structures", Engineer, 1935, 159, pp. 442-443
[3] Shepherd, R., "Strain Measurement Using Vibration Wire Gauges" Experimental Mechanics, 1964, 4(8), pp. 244-248
[4] Hornby, I., "The Vibrating Wire Strain Gauge", Strain Gauge Technology, Applied Science Publishers, 1982, pp. 327-348

[5] Clark, J., and Butler, J., "The Principles and Applications of Vibrating Wire Transducers", Report for Strainstall, 1979, pp. 1-13

[6] Thomas, D., Bromfield, M. and Evans, W., "Rapid Data Acquisition Using a Data Logger Interfaced to a Microcomputer", Strain, 1986, 22(4), pp. 183-189

9 Radial profilometry

J.A.Gilbert, P.Greguss, D.L.Lehner & J.L.Lindner

SUMMARY

This chapter describes a new device, called a radial profilometer, capable of contouring or measuring deflections on the inner surfaces of cavities found for example inside pipes, tubes and boreholes. The equations considered in designing and testing a prototype are described along with feasibility tests conducted to illustrate proof of principle. Future research is discussed, including a design for a fibre-based probe from which quantitative measurements are automatically extracted using a computer-based system.

INTRODUCTION

Profiling measurements are important in many fields, especially where chemical deposits cause corrosion, or combinations of thermal and mechanical stresses cause wear or produce cracks. Such conditions are typically encountered in nuclear power plants and in rocket engines where many components, designed to function at high temperatures and pressures, must be inspected periodically to avoid cataclysmic failures. Profiling techniques are also important in industrial manufacturing to identify parts outside of set tolerances, and to pinpoint defective or missing components.

Perhaps the oldest and best known method for profiling is photogrammetry where an object is photographed with one or more precision cameras from different incident angles and these photos are superimposed to form a three-dimensional image of the object. Deflections experienced by the test object are determined by measuring shifts in the position of its image relative to a known reference point [1]. An alternative approach for profiling is the shadow Moiré method in which a reference grating is placed in front of the test surface. When illuminated at an oblique angle, the reference grating casts a shadow on the surface, and the shadow deforms as the surface moves relative to the reference grating. The lines of the reference grating superimpose with those in the shadow to form a Moiré which can be analyzed to determine surface deflection [2].

Improvements have been made in both of these techniques by incorporating optical and image processing equipment for data acquisition and analysis. For example, Kahn-Jetter and Chu recently introduced a technique based on photogrammetry that combines stereoscopic analysis and digital image correlation techniques [3]. Two sets of stereo images are recorded for an undeformed and deformed test surface painted white and splattered with black paint. Digital image correlation techniques are applied to determine displacement. Sang, Lee and Chang developed a simple method based on shadow Moiré for measuring local buckling of thin walled members by projecting a pattern of parallel equally spaced lines on a test surface [4]. These lines are tracked by a computer and changes in their locations are related to movements in the out-of-plane direction. Gilbert et al combined the simplicity of shadow Moiré and the sophistication of photogrammetry in a technique called shadow speckle metrology in which random speckles, projected onto the test surface, are numerically correlated by a computer to profile structural components [5,6]. Optical fibres were later incorporated into shadow speckle metrology to facilitate remote testing [7]. Further improvements were made by Lehner et al for studying dynamic behaviour by introducing a technique called line broadening [8]. In that technique, a line is projected at an oblique angle onto a vibrating surface and the resulting image is tracked by a computer. Broadening takes place in areas where motion occurs and quantitative measurements are made to determine displacements.

The profiling techniques described above have been applied mainly to profile the outer surface of structural components. In general, these measurement systems rely on a large number of optical and electronic components, many of which are difficult to miniaturize. Although optical fibres offer potential for miniaturization and could be incorporated into such systems to profile remote surfaces, these systems (as well as the more conventional non-fibre-based systems) suffer from a limited field of view. This presents a problem, for example, when profiling the inner wall of a pipe. In this case, the imaging device would have to be translated along, and rotated around, the optical axis of the device to examine all points on the inner surface of the pipe; constraints which severely limit functional and real-time capabilities.

Ideally, a device for profiling the inner surface of a cavity should be rugged, compact, and capable of obtaining an unobstructed, complete, and comprehensive image of the cavity space in every direction. The measurement process should be sophisticated enough to fully characterize the test surface, yet simple enough to allow a technician to make the required measurements easily.

Unfortunately, it is virtually impossible to develop a practical device capable of recording a sphere of vision. However, most cavities can be regarded as cylindrical rather than spherical and information can be transformed, using stretching methods, onto a flat surface creating a 2-D representation of the 3-D cylindrical surface. This phenomenon called Flat Cylinder Perspective (FCP) forms the basis for a panoramic doughnut lens developed by Greguss at the Applied Biophysics Laboratory in Budapest, Hungary [9]. The lens produces an annular FCP image where the width of annular image corresponds to the vertical viewing angle; each concentric ring in the image plane is the loci of points recorded at a constant horizontal field angle [10]. The lens has been used to make holo-interferometric and speckle photographic recordings [11] and, when combined with the line broadening technique [8], can be used for radial profilometry.

ANALYSIS

Figure 9.1 shows a cross-sectional view of a device built to demonstrate proof of principle for radial profilometry. An unexpanded beam produced by a laser {1} passes through a 90 degree prism {2} (mounted on a transparent glass plate {4}) and reflects off a rotating front surface mirror {3}. The laser beam reflects off the mirror, passes through the glass plate {4}, and illuminates the interior wall of a cavity {5}. The figure shows a pipe being tested with its longitudinal axis positioned symmetrically with respect to the optical axis of the profilometer. In this case, the laser traces out a circular ring on the inside wall of the pipe. The image of the illuminated surface is captured by the panoramic doughnut lens {6}, and is projected by the projection lens {7} onto the image plane of a conventional 35mm camera {8}.

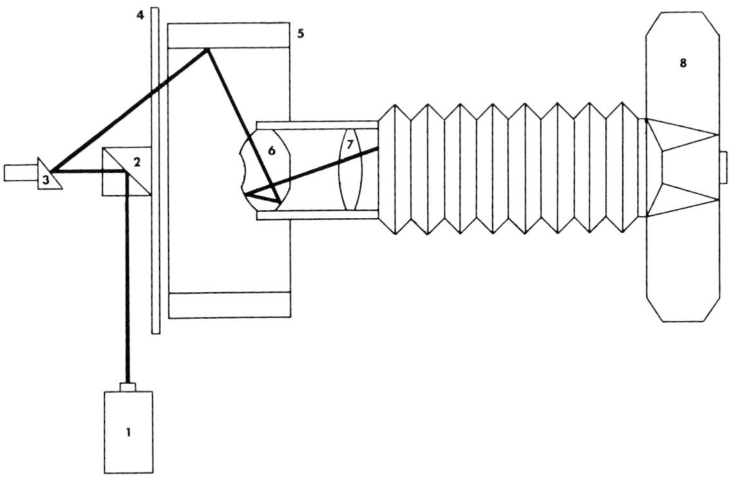

Figure 9.1: *A prototype designed and built to demonstrate proof of principle. 1. Laser; 2. Prism; 3. Rotating Mirror; 4. Glass Mounting Plate; 5. Test Object; 6. Panoramic Doughnut Lens; 7. Projection Lens; 8. 35mm Camera Body*

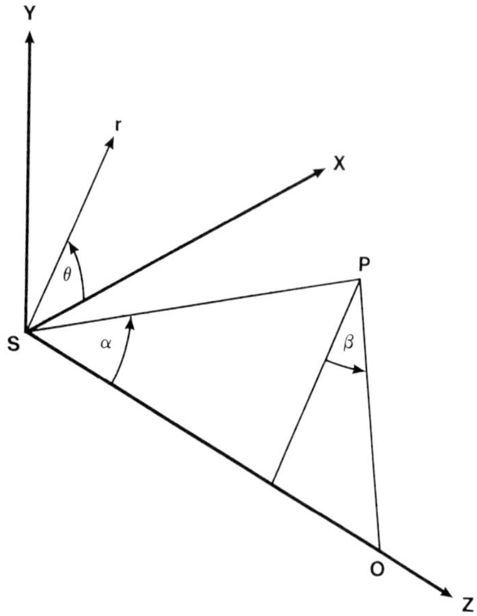

Figure 9.2: Coordinate axes systems used in the analysis

Figure 9.2 defines the cartesian and cylindrical coordinate systems to be used for subsequent analysis. In both systems, the optical axis of the profilometer lies along the z-direction; point S corresponds to the position of the rotating mirror denoted by {3} on Figure 9.1. Light is projected to point P at an angle α, and the image of the illuminated surface is captured at an angle β, measured with respect to a radial line lying in the r-z plane. When the surface moves relative to the optical axis, the projected image appears to shift along the z-direction through a displacement w. The corresponding radial displacement, r, is given by

$$r = \frac{w}{\tan\,[90 - \alpha] + \tan\beta} \tag{9.1}$$

The relationship between r and w is nonlinear, since β varies from point to point.

The displacement along the z-direction is mapped into the image plane by the panoramic doughnut lens via a mapping function, f, as follows

$$r' = f[w] \tag{9.2}$$

This function takes into account the magnification factor, and includes the FCP stretching methods used to create a 2-D representation of the 3-D cylindrical surface. Ideally, one would like to design a radial profilometer so that the nonlinear effects inherent in Equations (9.1) and (9.2) compensate one another over as large a range as possible, such that

$$r' = Cr + b \tag{9.3}$$

where C and b are constants.

By measuring r' and knowing C and b, Equation (9.3) can be solved for r. Once r is established for a known value of θ, the z-coordinate of the illuminated point can be calculated using

$$z = \frac{r}{\sin \alpha} \tag{9.4}$$

Therefore, one procedure for profiling a cavity is to initially establish a fixed global axis system in space with its z-axis aligned with that of the local system shown in Figure 9.2, and then to move the profilometer along the z-direction. The coordinates for r and θ are the same for the local and global systems; z-coordinates in the global system are calculated by taking into account the relative position of the local system.

EXPERIMENTAL

A radial profilometer was designed and built to satisfy the condition in Equation (9.3). The device was tested as shown in Figure 9.1 by inserting it into a circular pipe with an inner radius, R, equal to 2.25" (57.2mm). The optical axis of the profilometer was positioned parallel to the longitudinal axis of the pipe; the coordinate system shown in Figure 9.2 was used for analysis.

Figure 9.3 shows the image photographed when the radial profilometer is positioned at the centre of the pipe. The relatively thick circle is the laser trace; thin equispaced lines were drawn on the interior surface of the pipe to visually illustrate the nonlinear mapping inherent in Equation (9.2).

Figure 9.4 was recorded after the pipe was translated along the x-direction through a displacement, u, of 0.35" (8.9mm). Radial lines, drawn every five degrees, were superimposed on the photograph to aid in calibrating the profilometer. With θ measured from x,

$$r = [u^2 + R^2 + 2uR\cos(\theta + \gamma)]^{1/2} \tag{9.5}$$

where R is the inner radius of the pipe and $\gamma = \sin^{-1}[(u \sin\theta)/R]$.

Equation (9.5) can be derived using simple geometry, and defines the radial distance between the pipe and the optical axis of the profilometer for any angle θ. More importantly,

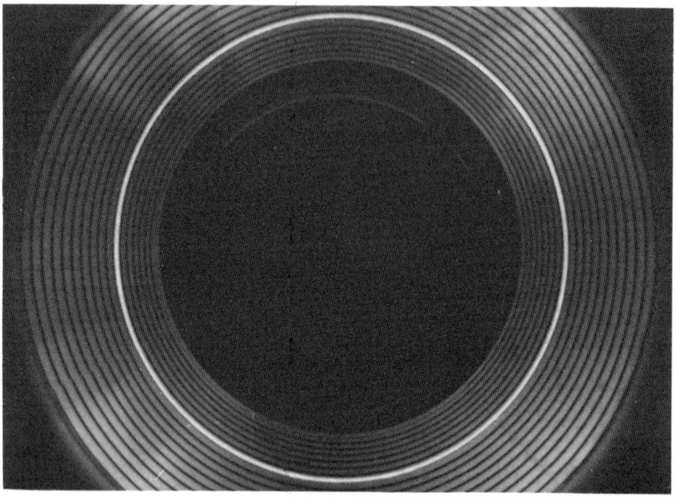

Figure 9.3: *Image of the test part with the pipe located symmetrically with respect to the optical axis of the profilometer. The thick white line is the laser trace. Thinner dark lines were drawn equally spaced along the longitudinal axis of the pipe and illustrate the nonlinear mapping of the PDL*

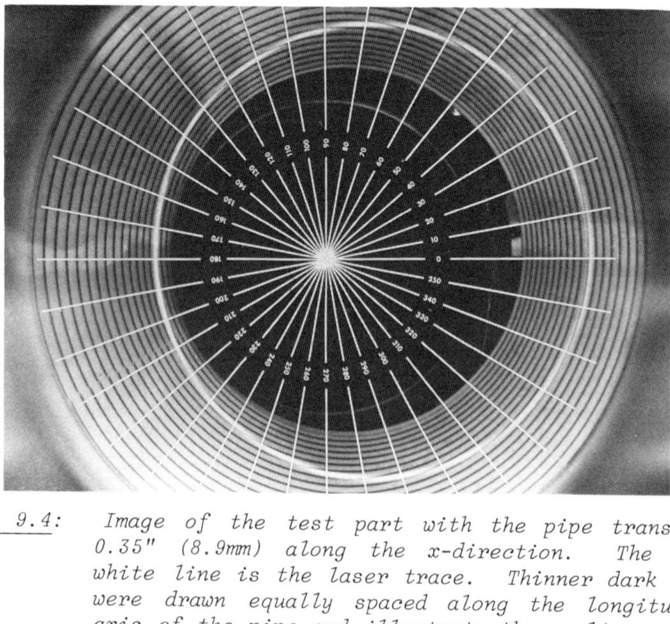

Figure 9.4: *Image of the test part with the pipe translated 0.35" (8.9mm) along the x-direction. The thick white line is the laser trace. Thinner dark lines were drawn equally spaced along the longitudinal axis of the pipe and illustrate the nonlinear mapping of the PDL*

the equation holds for any z-coordinate, since the optical axis of the device remains parallel to the longitudinal axis of the pipe and the cross section of the pipe is constant. This makes the translated pipe ideal for calibrating the profilometer, since r' can be measured on Figure 9.4.

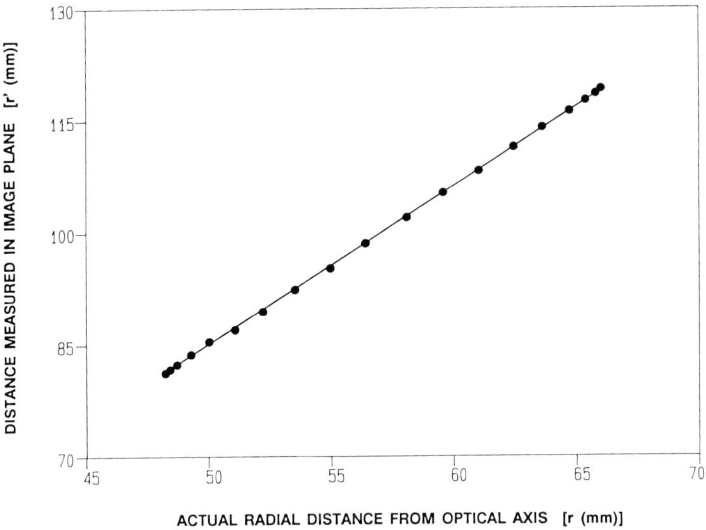

Figure 9.5: *Calibration curve for the profilometer for displacements rangings from 0 to 0.35" (8.9mm)*

Figure 9.5 shows the calibration curve established by plotting r [calculated on the basis of Equation (9.5)] versus r' [measured along the radial lines superimposed on Figure (9.4)] for points taken at five degree increments as θ ranged from 0 to 360 degrees. The curve holds over a 0.70" (17.8mm) range where the value of r lies between 1.9" (48.26mm) and 2.6" (66.04mm). In this range, the response is linear and governed by Equation (9.3) with C equal to 2.06. A value of b = -0.71" (-18mm) is established by interpolating the curve back to r = 0. No physical interpretation can be associated with this value of b (r' at r = 0), since the central portion of the PDL contains no image. It is simply one of the parameters required in Equation (9.3) to evaluate r within the calibrated range.

Further tests were conducted for several translations along the x-direction, ranging in magnitude from 0 to 0.35" (8.9mm). Figure 9.6 shows a plot of displacements observed on the micrometer stage used to translate the pipe relative to the device versus experimental data taken using the profilometer and Equation (9.3) for $\theta = 0$. The solid line shows the "ideal" response.

Finally, Figure 9.7 shows the result of placing several in-

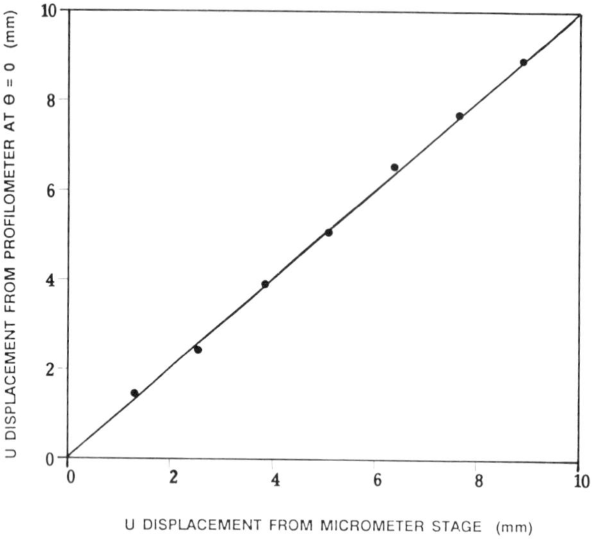

Figure 9.6: *Results for several translations along the x-direction ranging between 0 and 0.35" (8.9mm). Experimental data was taken at θ = 0, and evaluated using Equation (9.3)*

clusions of constant cross section on the inner wall of the pipe. In this case, the laser maps out shapes in the image plane which are "similar" to those of the inclusions. Each shape is reduced (or magnified) in size as defined by the constant, C, in Equation (9.3).

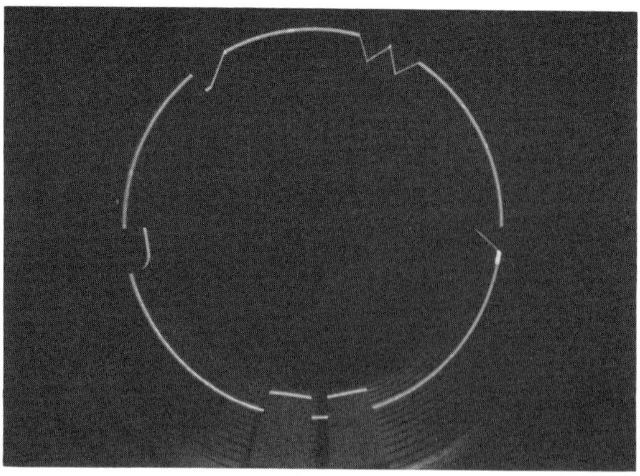

Figure 9.7: *Laser trace obtained when inclusions are placed on the inner wall of the pipe*

DISCUSSION

The device shown in Figure 9.1 has some major drawbacks which limit its potential for practical application. A practical device should be packaged so that it can be easily manipulated throughout a cavity. Analysis should be automated with a minimal amount of response time.

Figure 9.8 is a schematic of one of many optical configurations currently being evaluated for radial profilometry. The radial profilometer is shown inserted into a cylindrical cavity.

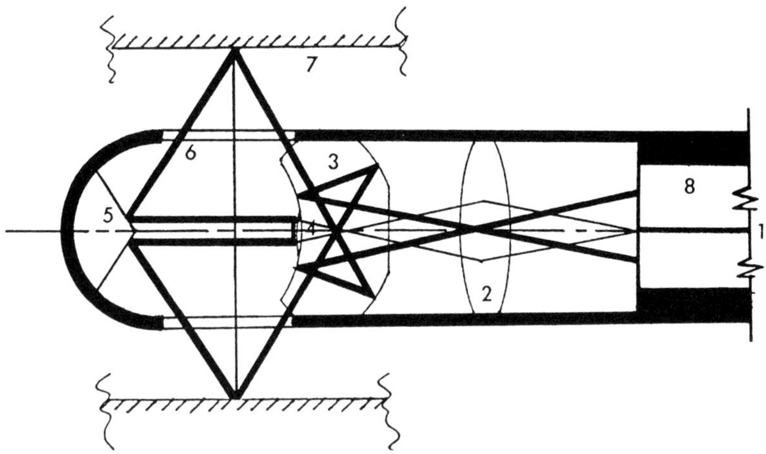

Figure 9.8: *One of the optical configurations currently being evaluated for radial profilometry. 1. Illuminating Fibre from Laser; 2. Projection Lens; 3. Panoramic Doughnut Lens; 4. Collimating Lens with Mask; 5. Conical Mirror; 6. Transparent Window; 7. Test Object; 8. Image Bundle to Computer*

A diverging laser beam (shown launched from a fibre optic labeled 1) is directed through a projection lens {2}. The beam passes through a panoramic doughnut lens {3} and is collimated and shaped by an appropriately masked collimating lens {4} to produce a thin ring. The ring reflects off a conical mirror {5} and passes through a transparent window {6} onto the test surface {7}. The image of the illuminated surface is captured through the transparent window {6} by the panoramic doughnut lens {3} and is projected by the projection lens {2} onto a coherent optical fibre bundle {8}. The bundle further transmits the image from the device to a photoelectronic-numerical processing system where changes in the image are recorded and analyzed using standard image acquisition and computer processing techniques.

The advantages of this device are that it is simple and relatively inexpensive, it can be miniaturized, there are no moving parts, the image is continuously displayed in the image plane, and measurements are completely automated. Future publications will discuss the advantages and disadvantages of this and other configurations proposed for radial profilometry. The range and accuracy for each configuration will be presented. Examples of practical applications will be given along with appropriate discussions of the numerical algorithms and computer software used to automatically extract quantitative measurements from the observed shifts in the position of the image.

CONCLUSION

This chapter has described a new device, called a radial profilometer, capable of contouring or measuring deflections on the inner surfaces of cavities found for example, inside pipes, tubes, and boreholes. An analysis has been presented which demonstrates that an entire cavity can be profiled simply by moving such a device through the cavity. Feasibility tests have shown that the profilometer can be designed so that information is extracted based on a linear calibration curve. Guidelines have been established for further development including automated analysis, and one of the designs currently under evaluation for practical implementation has been described in detail.

ACKNOWLEDGEMENTS

The authors wish to acknowledge support provided by the Center for Applied Optics at the University of Alabama in Huntsville and the Applied Biophysics Laboratory at the Technical University Budapest. Mr. Lehner and Mr. Lindner are funded as research assistants by the US Army Research Office in Research Triangle Park, NC under contract number DAAL 03-86-K-0014.

REFERENCES

[1] Gates, T.S.,"Photogrammetry as a means of mapping postbuckled composite surfaces" Experimental Techniques, 10(6), (1986), pp. 18-22

[2] Pirodda, L., "Shadow and projection moiré techniques for absolute or relative mapping of surface shapes", Optical Engineering, 21(4), (1982), pp. 640-649

[3] Kahn-Jetter, Z.L., Chu, T.C., "The use of digital image correlation techniques and stereoscopic analysis for the measurement of deformation on a warped surface", Proceedings of the SEM Spring Conference on Experimental Mechanics, New Orleans, June 8-13, 1986, pp. 820-826

[4] Sang, Z.T., Lee, G.C., Chang, K.C., "A simple method for measuring local buckling of thin-walled steel members", Proc. of the SEM Spring Conference on Experimental Mechanics, New Orleans, June 8-13, 1986, pp. 464-469

[5] Gilbert, J.A., Matthys, D.R., Taher, M.A., Petersen, M.E., "Shadow speckle metrology", Applied Optics, 25(2), (1986), pp. 189-203

[6] Gilbert, J.A., Matthys, D.R., Taher, M.A., Petersen, M.E., "Profiling structural members using shadow speckle metrology", accepted for publication in Optics and Lasers in Engineering, 1987

[7] Gilbert, J.A., Matthys, D.R., Taher, M.A., Petersen, M.E., "Remote shadow speckle metrology", Proceedings of the SEM Spring Conference on Experimental Mechanics, New Orleans, June 8-13, 1986, pp. 814-818

[8] Lehner, D.L., Dudderar, T.D., Matthys, D.R., Gilbert, J.A., "Vibration analysis using shadow speckle metrology and line broadening techniques", Proceedings of the SEM Spring Conference on Experimental Mechanics, Houston, Texas, June 14-19, 1987

[9] Greguss, P., US Patent No. 4, 566, 763, 1984

[10] Greguss, P., "The tube peeper: a new concept in endoscopy" Optics and Laser Technology, 17, (1985), pp. 41-45

[11] Greguss, P., "Panoramic holocamera for tube and borehole inspection", International Seminar on Laser and Optoelectronic Technology in Industry - A State of Art Review, June 25-27, 1986, Xiamen, P.R.C.

10 A hybrid technique for high-temperature dynamic fracture analysis

A.S.Kobayashi & K.H.Yung

SUMMARY

A hybrid technique, which utilizes the measured impact load and transient crack tip opening displacement (CTOD) together with a dynamic finite element code, was developed for high temperature dynamic fracture analysis. The dynamic finite element code was driven in its generation mode by the crack extension history and the impact load which is transmitted to the heated specimen through a push rod. The latter was determined by the transient CTOD which was monitored by a laser interferometric strain gauge (LISG). The procedure was verified through room temperature dynamic fracture toughness testing of alumina single edge notched bend bars together with additional KRAK gauge and in-situ load transducer measurements.

INTRODUCTION

Static fracture toughness testing at elevated temperature in excess of 1000°C is complicated by the lack of appropriate experimental procedures for monitoring the inevitable stable crack growth which occurs prior to crack instability. The constant stress-intensity-factor specimen [1,2] circumvents this issue by eliminating its crack length dependency but such specimen will not yield a crack-length independent dynamic stress intensity factor. This difficulty is further compounded by the added requirement for monitoring the applied load history during the dynamic fracture process at elevated temperature.

The traditional visual observations and procedures involving probes for monitoring the crack length during rapid crack propagation at elevated temperature are not practical. As a non-contacting measurement procedure, the laser interferometric strain gauge (LISG), which was developed by Sharpe [3], provides the sensitivity and the automatic data processing capability [4] for dynamic fracture testing at elevated temperature. This procedure has been used successfully to monitor the crack mouth opening displacement (CMOD) in ceramics and ceramic composite three-point-bend specimens at room temperature through 1200°C [5,6].

With dynamic fracture testing at elevated temperature, the transient load and load point displacement measurements, which were used for data reduction in [7], do not account for outside of the furnace for the time lag of transmitting the load to the specimen as well as for the stress wave filtering in the load train and are thus suspect. This difficulty, however, can be circumvented using the hybrid experimental-numerical procedure where the load train is incorporated into the dynamic finite element model of the fracture specimen. Dynamic finite element analysis of the entire loading system, when properly conducted, will provide the correct dynamic state in the fracture specimen.

The purpose of this chapter is to present a hybrid experimental-numerical procedure for determining the dynamic fracture toughness of a fracture specimen at elevated temperature.

EXPERIMENTAL PROCEDURE

Laser interferometric strain gauge (LISG)

The laser interferometric strain gauge generates optical interference fringes from a coherent light reflected from two small reflective microhardness indentations as shown in Figure 10.1. The two fringe patterns are displaced relative to the fixed incident laser beam as the two indentations move relative to each other. Although rigid body translation in the specimen can be removed by a simple average of the two displacement records, excess rigid body rotation will cause the reflected beams to move apart in space without interference. Fortunately, the small specimen rotation associated with the brittle ceramic fracture specimens, precluded such occurrence but could pose a problem in testing ductile metallic specimens. The moving interferometric fringes were monitored by two photomultiplier tubes and recorded on a transient wave form recorder.

Loading system

The loading system consists of a drop weight tower, which is mounted integrally with the furnace, with a maximum operating temperature of 1500°C (Figure 10.2). The two port holes and one door, which are shielded with fused silicate glass panes to prevent the undesirable convection of air currents in the furnace chamber, provided the LISG access to the fracture specimen. The impact load is monitored by a load transducer at the top end of the push rod and outside of the furnace. An additional load transducer is placed between the bottom end of the push rod and the fracture specimen for room temperature testing. The load transducer outputs are recorded on a transient wave form recorder.

Specimen

The specimens used in this study are single-edge notched, three-point-bend, alumina (Coors AD995) specimens as shown in Figure 10.3. Since ceramics are translucent and do not reflect the

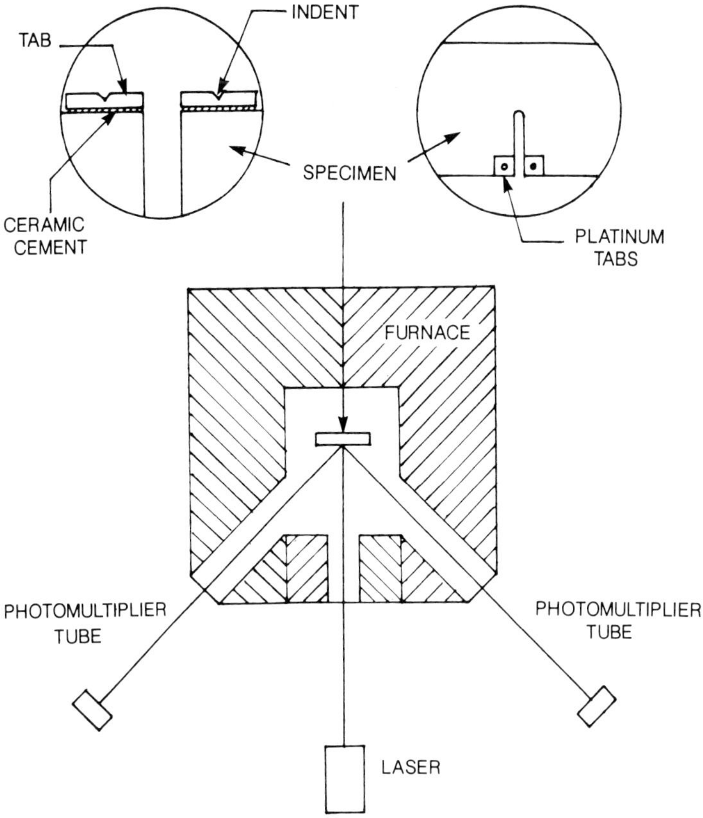

Figure 10.1: Schematic diagram of laser interferometric strain gauge (LISG)

laser beam, a process for attaching a reflecting surface, which reproducibly yielded distinct and bright interference fringes at elevated temperature, was developed. The process, which follows that of Sharpe [8], consists of bonding thin platinum tabs, typically 3.0 x 2.0 x 1.0mm, to the specimen surface with Aremco Ceramabond 569 adhesive. An alumina slurry was used to polish the tabs prior to bonding. The tabs, which were positioned next to the crack tip, are then indented by a Vicker's diamond indenter at a load of 500 grams to produce a uniform, symmetrical pyramidal indentations approximately 0.08mm in diagonal length. Each indent was located 0.4mm from the crack tip.

Four specimens were also instrumented with KRAK gauges* to

* TTI Division, Hartrum Corporation, St. Augustine, FL 32085, USA.

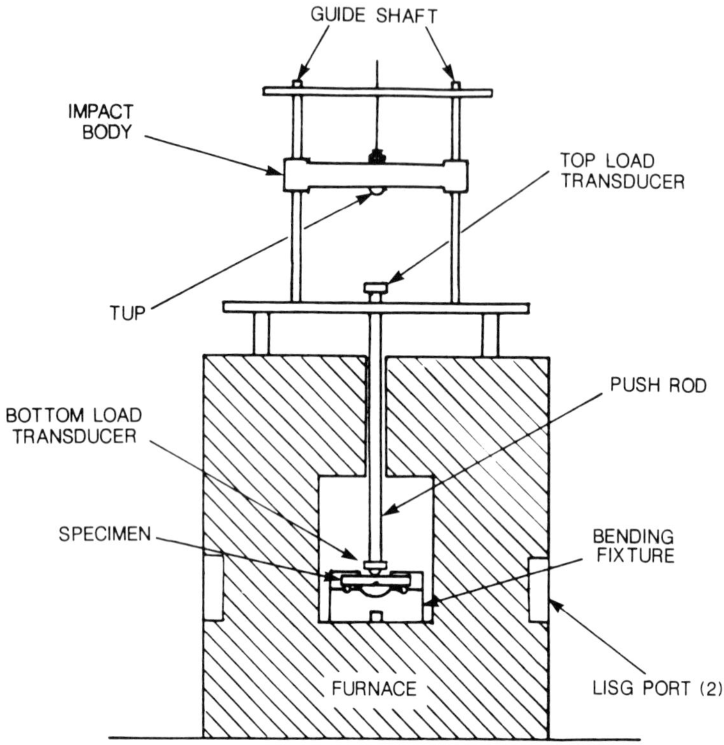

Figure 10.2: *Schematic diagram of drop-weight impact system and furnace with LISG ports*

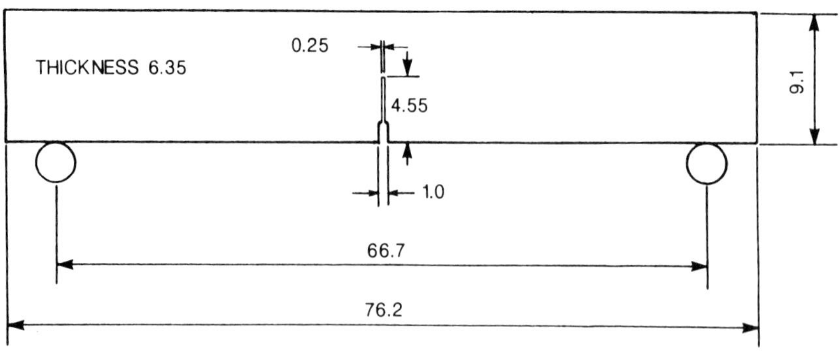

Figure 10.3: *Dimensions of specimen*

determine the crack extension history during rapid fracture at room temperature. This data was used to construct a master curve which related the measured CTOD between the two indentations and the measured instantaneous crack tip location during rapid crack propagation.

DYNAMIC FINITE ELEMENT ANALYSIS

An implicit dynamic finite element code was executed in its generation mode [9] to compute the dynamic initiation and propagation fracture toughnesses. Figure 10.4 shows the finite element breakdown of one half of the specimen and the push rod used in this analysis. The finite element model in this figure represents a transient state where the specimen is in contact with the two end supports and the push rod. In actuality, the specimen vibrates at nearly its natural frequency upon impact and thus these contacts are made and broken intermittently during the fracture process. As discussed in the next section, these variable boundary conditions must be incorporated into the preparation analysis [9] of the fracturing specimen. The input data to the finite element analysis thus consists of the crack extension history, which was determined by either the KRAK-gauge data at room temperature or by the CTOD data through a master curve at elevated temperature, and the measured impact load. The dynamic initiation stress intensity factor was computed by using a calibrated crack opening displacement (COD) procedure. The dynamic propagation stress intensity factor was determined from the numerically computed dynamic energy release rate using Freund's relation [10].

VERIFICATION OF TEST PROCEDURE

The accuracy of the test procedure was verified by room temperature testing of alumina fracture specimens. Additional data acquired through a second load transducer (Figure 10.2) at the impact point on the specimen and a KRAK gauge placed at the remaining ligament of the prenotched bar, were used to check the accuracy of the finite element modelling of the load train in Figure 10.4 and the master curve which relates the crack extension history with the CTOD data.

Figure 10.5 shows typical data from the top and bottom load transducers and the KRAK-gauge. Figure 10.6 shows the corresponding LISG data together with the corresponding KRAK-gauge data. The KRAK gauge data shows a definite crack acceleration phase which was also noted in previous studies [11,12,13]. Additionally, the slight ductility of the KRAK gauge causes the gauge to lag behind the actual crack tip in the brittle ceramics. This lag in the apparent crack tip location, as measured by the KRAK gauge, was noted by the unreasonably small dynamic initiation fracture toughness at the instant of crack propagation as determined by the KRAK gauge [14].

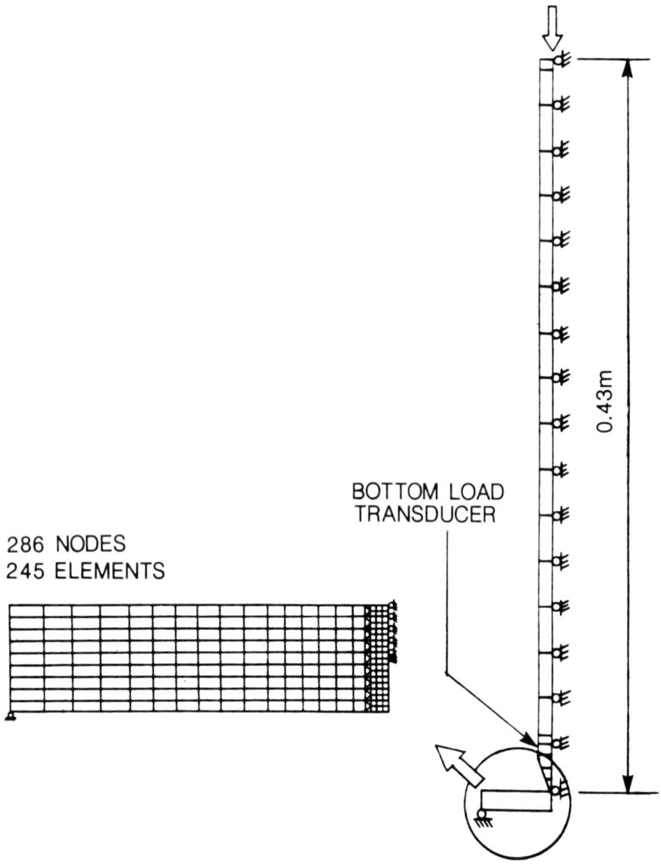

Figure 10.4: Finite element breakdown

The overall KRAK gauge response, with a time lag, should accurately represent the crack propagation history, while the LISG data, which records the CTOD prior to and after crack propagation, is free of such time lag and the two sets of data could be used to delineate this time lag by matching their corresponding crack extension rates. Figure 10.6 shows typical LISG and KRAK gauge data with the matched periods as indicated. The difference between the actual and apparent start of crack propagation represents the sought time lag. For a specimen of the same geometry, the resultant LISG fringe motion, i.e., the algebraic sum of the left and right LISG data, is uniquely related with the instantaneous crack length as shown in Figure 10.7. This fringe number versus CTOD master curve can be used to locate the instantaneous crack length from the LISG data provided a proportional adjustment is made to accommodate for the changes in elastic modulus at elevated temperature.

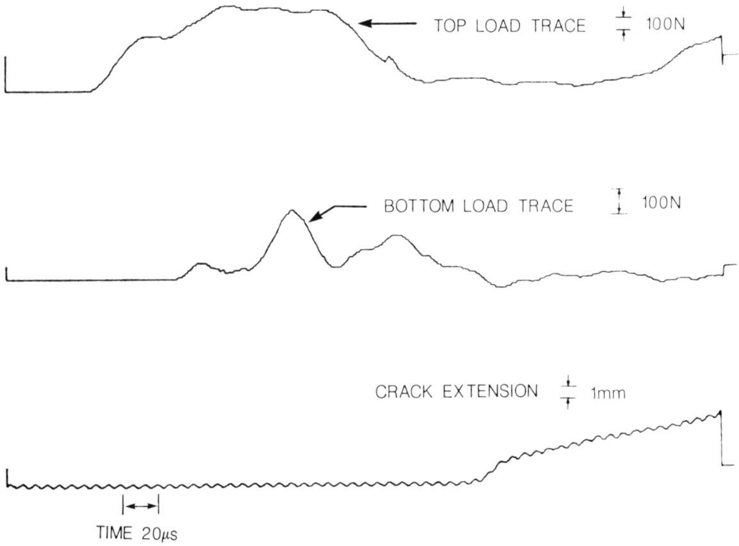

Figure 10.5: *Time-load traces for both load transducers and crack extension data. KH021087*

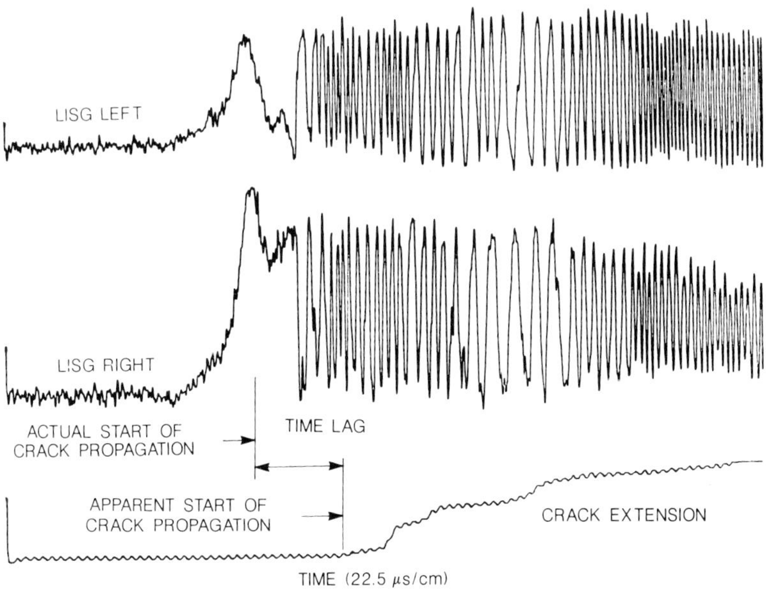

Figure 10.6: *LISG data for both PMTS and crack extension data KH021587*

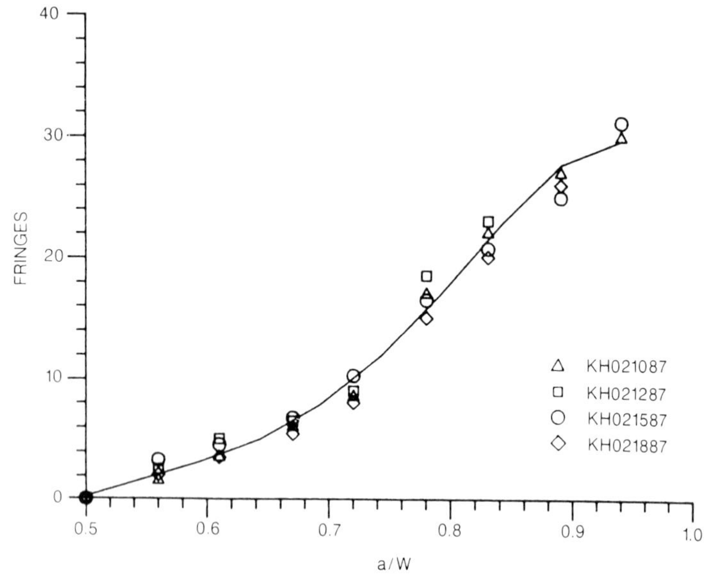

Figure 10.7: Crack extension versus CTOD master curve

The dynamic response of the finite element model of Figure 10.4 was checked by applying the recorded impact pulse at the top end of the push rod in the dynamic finite element model and computing the output load at the specimen or the bottom end of the push rod (Figure 10.8). Also shown in Figure 10.8 is the measured output load at room temperature by the bottom load transducer. A check of the finite element analysis showed that the divergence between the numerically determined and measured output loads after 125 microseconds of lapse time was attributed to the tensile stresses at the impact point after this moment and indicates a separation between the push rod and specimen at this instant. This separation must thus be incorporated in the dynamic finite element analysis.

The actual dynamic finite element analysis thus consists of two computations, one with the push rod connected to the specimen and the other with the computed output load, which is truncated by the generation of tensile stresses at the specimen end of the push rod, prescribed at the impact point of the specimen (Figure 10.7).

DYNAMIC FRACTURE TOUGHNESS OF ALUMINA

The above procedure was used to determine the dynamic initiation and propagation fracture toughness of the alumina, single-degree-notched three-point-bend specimens. Figure 10.9 shows the measured crack extension history which was then used together with

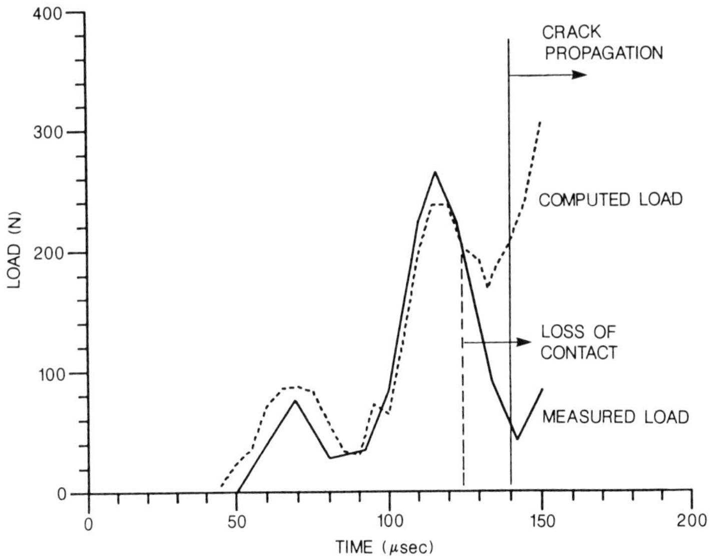

Figure 10.8: *Comparison of measured and computed load traces*

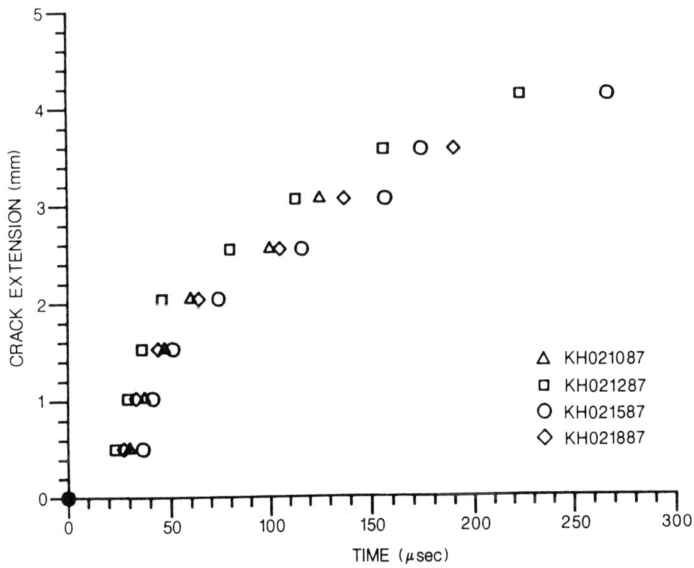

Figure 10.9: *Crack extension versus time relation*

the impact pulses (as recorded by the top load transducer) in the generation of dynamic finite element analysis. Figure 10.10 shows the variations in dynamic fracture toughness with crack extension at room temperature in the four specimens. Figure 10.11 shows the resultant crack velocity versus dynamic fracture toughness where a reasonable reproducibility in the data is noted. Without the cluster of data at the left end it is not difficult to imagine the existence of the well-known gamma shape curve which has been observed in metals and polymers [15,16]. This nontraditional crack velocity versus dynamic fracture toughness relation in ceramics is consistent with previous findings [11,12,17].

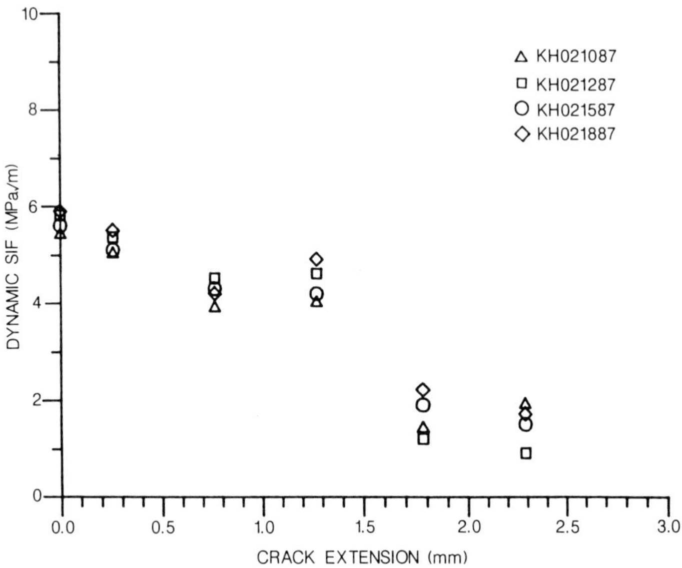

Figure 10.10: Dynamic SIF versus crack extension relation

CONCLUSIONS

A procedure suitable for dynamic fracture characteristics of structural ceramics at elevated temperature has been presented. This test procedure was used to determine the crack velocity versus dynamic fracture toughness relation of alumina at room temperature.

ACKNOWLEDGEMENT

The results reported here were generated under NASA contract NAGW-199.

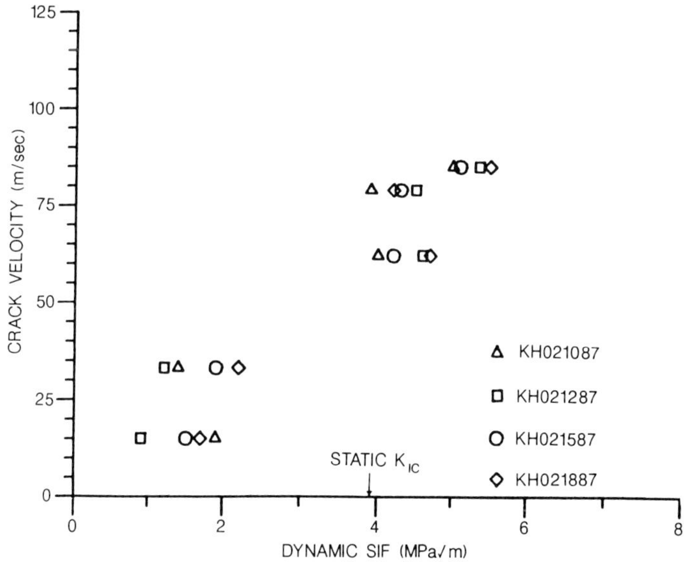

Figure 10.11: Dynamic SIF versus crack velocity relation

REFERENCES

[1] Evans, A.G., "A Simple Method for Evaluating Slow Crack Growth in Brittle Materials", International Journal of Fracture, Vol. 9, No. 3 (1973) pp. 267-275

[2] Kobayashi, A.S., Emery, A.F., Gorum, A.E., Basu, T., "Fracture Toughness of Alumina at Room Temperature to 1600°C", Proceedings of the Third International Conference on Mechanical Behaviour of Materials, ed. by K.J. Miller and R.F. Smith, Pergamon of Canada, Vol. 3 (1979) pp. 3-10

[3] Sharpe, W.N., Jr., "Interferometric Surface Strain Measuremont", International Journal of Nondestructive Testing, Vol. 3 (1979) pp. 59-76

[4] Hartman, G. and Nicholas, T., "An Enhanced Laser Interferometer for Precise Displacement Measurements", Experimental Techniques, Vol. 11, No. 2 (1987) pp. 24-26

[5] Jenkins, M.G., Kobayashi, A.S., Sakai, M., White, K.W. and Bradt, R.C., "Fracture Toughness Testing of Ceramics Using a Laser Interferometric Strain Gage", to be published in Bulletin of American Ceramic Society

[6] Jenkins, M.G., Kobayashi, A.S., White, K.W. and Bradt, R.C., "Crack Initiation and Arrest In a SiC Whisker/A1203 Matrix Composite", to be published in Journal of American Ceramic Society

[7] Gonczy, S.T. and Johnson, D.L., "Impact Fracture of Ceramics at High Temperature", Fracture Mechanics of Ceramics, Vol. 3, ed. by R.C. Bradt, D.P.H. Hasselman and F.F. Lange, Plenum Press, New York (1978), pp. 495-506

[8] Sharpe, W.N. Jr., "Preliminary Development of an Interferometric Strain Gage for use on Nosetip Materials Subjected to Thermal Shock", AFML/TR-76-63 (June 1976)

[9] Kobayashi, A.S., "Dynamic Fracture Analysis by Dynamic Finite Element Method - Generation and Propagation Analysis", Non-linear and Dynamic Fracture Mechanics, ed. by N. Perrone and S.N. Atluri, ASME AMD Vol. 35 (1979) pp. 19-36

[10] Freund, L.B., "Crack Propagation in an Elastic Solid Subjected to General Loading: II, Non-Uniform Rate of Extension", Journal of the Mechanics and Physics of Solids, Vol. 20 (1972) pp. 141-152

[11] Kobayashi, A.S., Emery, A.F. and Liaw, B.M., "Dynamic Fracture Toughness of Glass", Fracture Mechanics of Ceramics, Vol. 6, ed. R.C. Bradt, A.G. Evans, D.P.H. Hasselman and F.F. Lange, Plenum Press (1983) pp. 47-62

[12] Kobayashi, A.S., Emery, A.F. and Liaw, B.M., "Dynamic Fracture Toughness of Reaction Bonded Silicon Nitride", Journal of American Ceramic Society, Vol. 66, No. 2 (Feb. 1983) pp. 151-155

[13] Huang, J.L., Virkar, A.V. and Huber, R.J., "Fast Crack Propagation Across Interfaces", Fracture Mechanics of Ceramics, Vol. 6, ed. R.C. Bradt, A.G. Evans, D.P.H. Hasselman and F.F. Lange, Plenum Press (1983) pp. 121-136

[14] Kobayashi, A.S. and Yang, K.H.m "Dynamic Stress Intensity Factor Versus Crack Velocity Relation", Prepint of Japan-US Joint Seminar on Materials Needs for Severe Conditions, Tokyo Institute of Technology (May 1986) pp. 35-45

[15] Dally, J.W., "Dynamic Photoelastic Studies of Fracture", Experimental Mechanics, Vol. 19 (1979) pp. 349-361

[16] Kobayashi, A.S., Ramulu, M., Dadkhah, M.S., Yang, K.H. and Kang, B.S.-J., "Dynamic Fracture Toughness", International Journal of Fracture, Vol. 30 (1986) pp. 275-285

[17] Liaw, B.M., Kobayashi, A.S. and Emery, A.F., "Effect of Loading Rate on Dynamic Fracture of Reaction Bonded Silicon Nitride", Fracture, Seventeenth Volume, ASTM STP 905, ed. by J.H. Underwood, R. Chait, C.W. Smith, D.P. Wilhem, W. A. Andrews and J.C. Newman (1986) pp. 95-107

11 Strain distribution ahead of the crack front in SEN concrete beams in three-point bending

S.E.Swartz & T.M.E.Refai

SUMMARY

This chapter reports the results of tests on a small, unreinforced concrete beam with a single-edge-notch (SEN) in three-point bending in which electrical-resistance strain gauges were used to monitor strain distributions as the crack progressed through the beam. Comparisons with an elastic solution confirm the suitability of the methods of linear-elastic-fracture mechanics (LEFM) as a practical means of estimating the fracture parameter, critical crack-tip-opening-displacement ($CTOD_c$) even through a process zone of significant size exists ahead of the open crack.

NOTATION

a_o, a	notch depth, crack length
B,W,S	width, depth, span of beam
E_c, μ	modulus of elasticity, Poisson's Ratio
$f'_c, f'_t, \varepsilon_t$	compressive strength, tensile strength and cracking strain of concrete
K_I	mode one stress intensity factor
P,M	applied load and associated midspan moment
R	neutral axis depth ratio
r_p	process zone length
x, y r, θ	rectangular and polar coordinates
v	measured crack-mouth-opening displacement (CMOD)
z	thickness of knife edge supporting CMOD measuring transducer
δ, δ_c	crack-tip-opening-displacement (CTOD) and critical displacement ($CTOD_c$)
$\varepsilon_y = \varepsilon_o$	longitudinal strain at cracked section
ζ	$= -0.065\,(1 - a/W)^2 - 3.483\,(1 - a/W) - 0.120 + 5.706\,(1 - a/W)^{-1} + 0.166\,(1 - a/W)^{-2}$
η	distance from open crack tip to end of measured crack, see Figure 11.2
$\sigma_r, \sigma_\theta, \tau_{r\theta}$	normal and shearing coordinate stresses
$\sigma_y = \sigma_o$	normal longitudinal stress at cracked section
ξ	$= W - a$

INTRODUCTION

In the linear-elastic-fracture mechanics (LEFM) analysis of a propagating crack in a single-edge-notch (SEN) beam in three-point bending the analytical solution for the stress (and strain) distribution ahead of the crack may be obtained in a number of ways. For a given crack length the stress in the region of the crack tip is related to the opening-mode, stress-intensity-factor K_I. The crack will propagate when K_I reaches a critical value K_{IC} which is presumed to be a material property which is independent of crack length and specimen size.

In attempting to apply this concept to concrete, problems arise due to the variability in results which appear to be crack length and beam size dependent. A number of methods have been proposed to overcome these problems while still using LEFM concepts. These are summarized in Reference [1] in which it is seen that valid K_{IC} estimates may be obtained if sufficiently large beams are used, along with precracking and reasonable methods to estimate the crack length. It is also noted that a fairly large "process zone" of microcracking precedes the crack front.

Another approach which may be useful is to determine the critical crack-tip-opening displacement ($CTOD_c$) and use this as the fracture parameter to describe the process of crack growth [2,3]. A method proposed by Jenq and Shah [4] uses both K_{IC} and $CTOD_c$ in a non-linear fracture model although both parameters are determined using LEFM.

To the writers' knowledge, no attempt has been made to obtain experimental verification of any analytical solution for the stress distribution in a concrete beam ahead of the crack front other than that presented by Rood [5].

The objectives of the present work are: 1. to obtain reliable strain data in a concrete beam in the region ahead of the crack; 2. to compare the measured strain distributions with those predicted by an LEFM formula; 3. to calculate the $CTOD_c$ using a simple formula based on LEFM along with beam test data and verify its independence with crack length.

THEORETICAL BACKGROUND

The subject beam geometry is shown in Figure 11.1 in which it is to be noted that the crack length a is different from notch depth a_o. It is also noted that a displacement is to be measured across the notch/crack at the surface of the beam. This is denoted the CMOD (crack-mouth-opening-displacement) and is identified by the symbol v.

In Figure 11.2 the notch-crack-process zone geometry is described following ideas presented in References [2] and [6]. In this the distance from the edge of the open crack, to the measured crack length a is η. At this location the origin of coordinates is placed. Note that the length η is not necessarily the length of the process zone r_p.

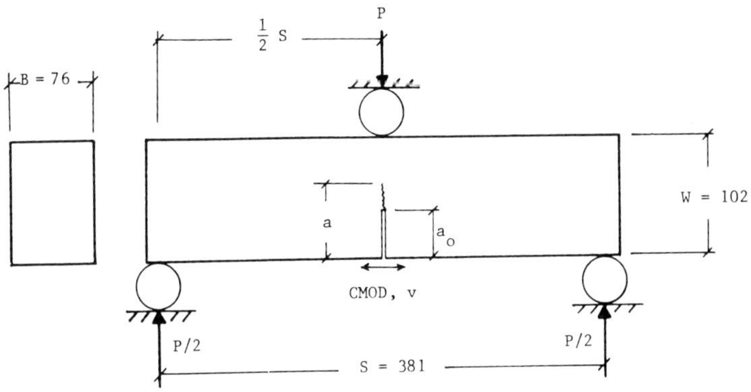

Figure 11.1: Beam in three-point bending

When the crack front is identified, for instance by the use of dye (Figure 11.3) the parameter η is the distance from the edge of the front to the average crack depth. The measurement is made at the centre of the cross-section as the CMOD is also measured at that location. It is seen that at any location above the crack front $\delta < \delta_c$ since the dye has not penetrated. It is assumed here that at the crack front $\delta = \delta_c$.

Based upon this simplified geometry, an estimate of δ_c is given by

$$\delta_c = \frac{R(W - a) + \eta}{R(W - a) + a + z} v \qquad (11.1)$$

It is assumed (and must be verified) that the neutral axis ratio R is constant for different crack lengths. The terms η and v are determined from test data as described in the sequel.

To determine R, use is made of a stress function (following Williams[7]) of the form

$$\sigma_o = -a_1 r^{-1/2} + 3\, a_3\, r^{1/2} - 5\, a_5\, r^{3/2} + \ldots \qquad (11.2)$$

in which the first three coefficients can be obtained explicitly.

From LEFM, the opening-mode, stress-intensity factor K_I is defined as

$$K_I = \lim_{r \to 0} \sigma_\theta(r, \theta=0) \sqrt{2\pi r} \qquad (11.3)$$

By applying this to Equation 11.1 it is seen that

$$a_1 = -\frac{K_I}{\sqrt{2\pi}} \qquad (11.4)$$

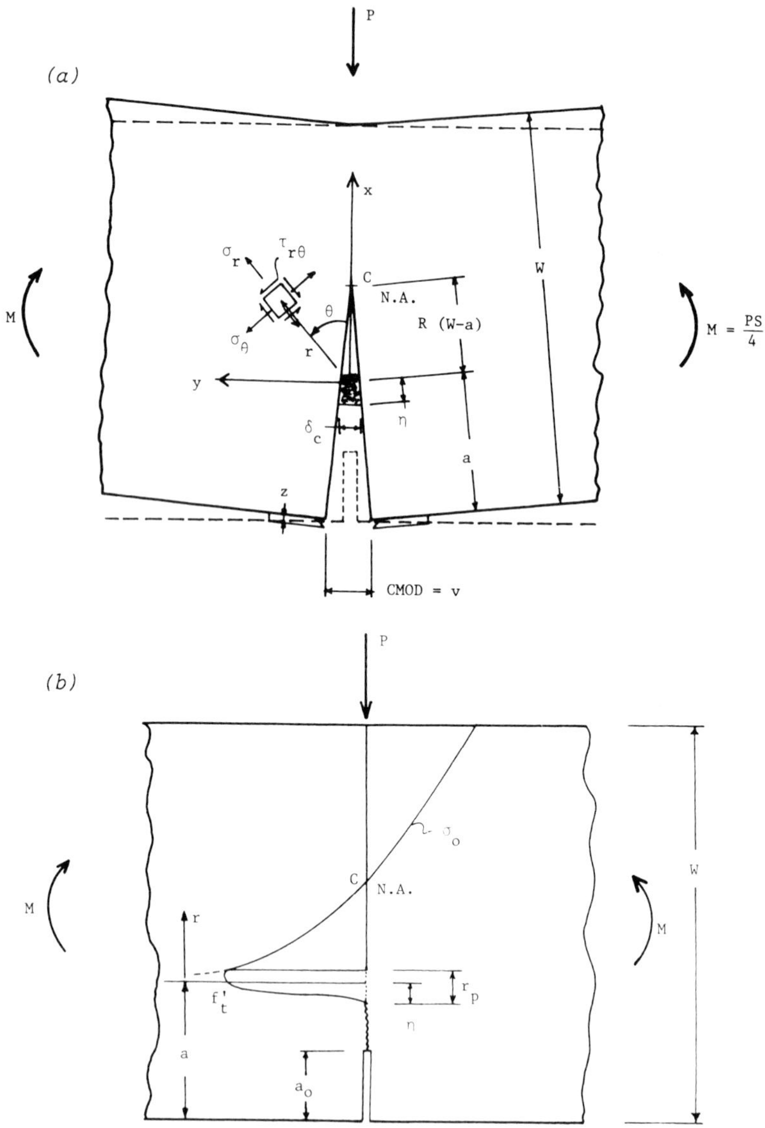

Figure 11.2: *(a) Crack geometry, SEN beam with crack*
(b) Stress distribution, SEN beam with crack

By use of boundary collocation in conjunction with the Williams stress function for beams with the geometry given in Figure 11.1 (e.g. $\frac{S}{W}$ = 3.75), K_I is given by [8].

$$K_I = \frac{P\ S}{4\ BW^{1.5}} \zeta(\frac{a}{W}) \qquad (11.5)$$

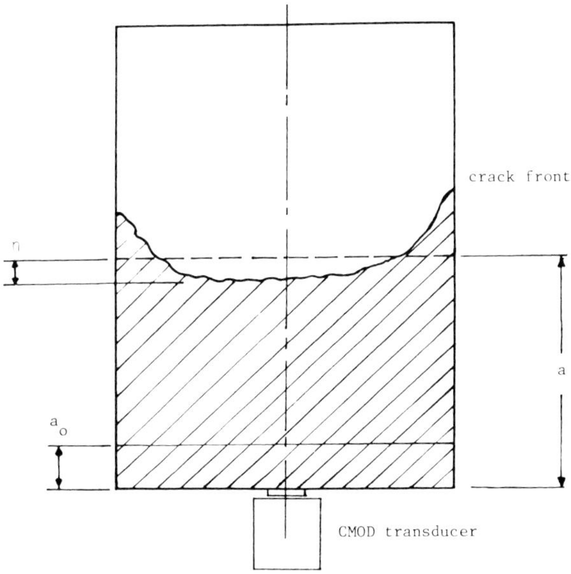

Figure 11.3: Typical crack appearance, 4 in. deep, SEN beams

To determine a_3, a_5 the equilibrium of internal and external forces and moments is considered which yields

$$a_3 = \frac{\frac{16}{21} a_1 \xi^{3/2} + \frac{P S}{4 B}}{\frac{8}{35} \xi^{5/2}} \qquad (11.6)$$

$$a_5 = \frac{\frac{8}{15} a_1 \xi^{3/2} + \frac{P S}{4 B}}{\frac{8}{35} \xi^{7/2}} \qquad (11.7)$$

Then, R is determined from the requirement that $\sigma_\theta = \sigma_0 = 0$ at the location of the neutral axis. This gives

$$R = \frac{1}{10 a_5 \xi} (3 a_3 + \sqrt{9 a_3^2 - 20 a_1 a_5}) \qquad (11.8)$$

In these equations, $\xi = W - a$ and ζ is given by the function listed in the Notation [8].

EXPERIMENT

Strain measurement

The first writer attempted to measure the strain distribution ahead of the crack front through use of photoelastic coatings [9]. This was unsatisfactory as the coating reinforced the beam and influenced the fracture process.

Furthermore, the coating itself cracked and although the crack in the coating appeared to follow the crack in the beam some uncertainty existed as to the reliability of the fringe pattern. One thing the pattern did indicate was relatively low strain concentrations under the load-point roller. Actually, considering the load level, roller diameter, and beam width, contact stresses should be low. The roller diameter was 51mm and beam width was 76mm.

Because of this experience the use of photoelasticity was abandoned. Other optical approaches might be suitable, e.g., holographic interferometry or Moiré. The latter approach has been used by Cedolin et al [10] who applied this to tension specimens with and without notches. The fringe patterns using a grid spacing of 1000 lines/mm were converted into strain and displacement distributions.

In the work reported here, electrical-resistance strain gauges were bonded to the concrete beam shown in Figure 11.4. One problem with using strain gauges on concrete is the variation in response caused by the possible presence of large, aggregate particles in the immediate region of the gauge. The maximum aggregate size used in the mix was 19mm. If the usual rule of thumb is followed in which the strain gauge length is three times the size of the maximum aggregate then a gauge length of 57mm is indicated.

On the other hand, the steep strain gradients and desire to obtain data at closely-spaced points dictated that short gauge lengths be used. Thus, gauges with an effective length of 1.5mm were selected and the concrete surface prepared to minimize the effects of varying aggregate size.

As shown in Figure 11.4, twenty-two gauges were placed (eleven per side) in a region about 60mm deep at a spacing of about 5mm centre to centre. These gauges were connected with three lead wires to an OPTIM (OPTILOG) data acquisition system in a quarter bridge circuit.

Excitation was by twin, constant current at 13.9mA in an "integrating" mode in which the samples are integrated over the period of the AC mains (1/60 sec). In this mode the maximum speed is 40 samples per second. The sensitivity scale selected was 0.3μm/m. The system was driven by an Apple II computer.

Test specimens

Forty beam specimens of the geometry shown in Figure 11.1 were constructed as part of an extensive testing programme [5] of which the tests reported here were a small part. Data from these other beams were necessary however to supplement the data

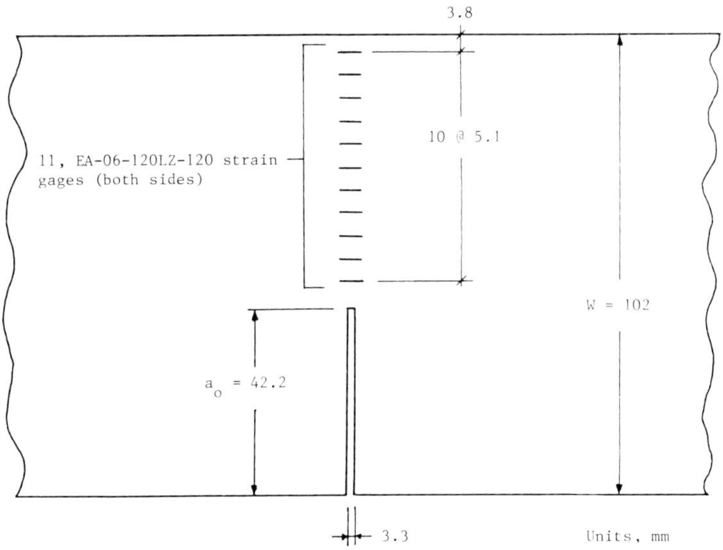

Figure 11.4: Strain gauge locations on tested beam

from the strain-gauge beam. The mix design is given in Table 11.1. The beams were removed from their moulds after one day and placed in a 100% humidity curing room until about one week prior to testing. They were then removed from the curing room and air-dryed. The age of most of the beams at time of testing was between 138 and 145 days. The beam for which results are reported here had a test age of 450 days. However, it is expected that strength gain was very little for the additional 300 plus days.

Table 11.1

Mix design

Water/cement	0.50
Sand fineness modulus	2.91
Maximum aggregate size	19mm
% water by weight	6.62
% type I cement by weight (SG = 3.15)	13.24
% sand by weight (SG = 2.65)	32.68
% aggregate by weight (SG = 2.56)	47.46
Unit weight of mix	2400 kg/m^3
Compressive strength, f_c'	55.4 MPa
Tensile strength, f_t' (split cylinder)	4.3 MPa
Tensile strain (est), ε_t	150 m/m
Modulus of elasticity	36.8 GPA
Poisson's ratio (est)	0.18

Testing procedure

Twenty-eight beams were precracked and dye inserted to reveal the crack front following procedures described in [5] and [11]. The crack front shown in Figure 11.3 is representative of many of the beams. These beams were used here only for the purposes of obtaining an average value of η and for compliance calibration to estimate crack lengths.

The instrumented beam was tested using an MTS closed-loop system operating in strain control (based on the output of the CMOD gauge). Prior to loading, the beam was notched to a depth a_0 = 41.4mm. The loading cycles are displayed in Figure 11.5. For each cycle of loading after the first, the average crack length was estimated from the compliance of the slope obtained after crack closure is overcome using the procedure described in [5] and [11]. The compliance calibration curve was established from the earlier tests on the twenty-eight beams.

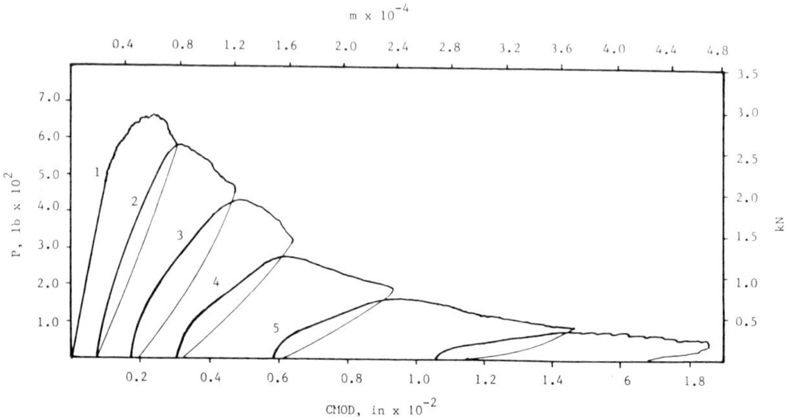

Figure 11.5: Load cycles, tested beam

Strain values were obtained for a given crack length by loading the beam to a level well below the maximum for that crack length. At the same time, the attempt was made to obtain data at about the same load for each crack length. This was achieved for all but cycles 5 and 6.

Difficulties were encountered in zeroing some of the gauges and three were abandoned - thus only 19 channels of data were obtained. In general we have experienced problems using the data acquisition system with three wire leads. The system appears to be very intolerant of slight resistance deviations from 120 ohms. Problems were also encountered with zero shifts even though the gauges were load cycled to the maximum strain levels expected to be encountered. Thus, in all cases, zero readings were taken before and after the load data were obtained and the latter values used to correct the readings at load. These zero shifts cannot be attributed to time effects as the load was applied only for about 60 seconds. They might be due to material non-linearity and crack-closure. As the load-displacement diagram (Figure 11. 5) shows, zero shifts for cycles 1 and 2 were not measurable and were fairly small for cycles 3 through 5.

TEST RESULTS

Plots of average strain at each station (where available) are presented in Figure 11.6 for each of the first 5 cycles of load. By the sixth cycle the crack length was so great (a = 100.6mm or $\frac{a}{W}$ = 0.99) that it was not possible to obtain strain data. Also plotted in Figure 11.6 are strain distributions based upon Equation 11.2 as follows:

$$\varepsilon_0 = \frac{1}{E_c}(\sigma_o - \mu\sigma_r) \qquad (11.9)$$

in which μ = 0.18. The contribution of σ_r is ignored here and

$$\varepsilon_o \approx \frac{\sigma_o}{E_c} \qquad (11.10)$$

From the plots presented in Figure 11.6 it is seen good agreement is obtained between the experimental and theoretical strains except for cycle 5 and for which $\frac{a}{W}$ = 0.78 and the distance from the end of the average crack to the load point is about 22mm.

The values of R obtained from Equation 11.8 and those obtained from the plots of experimental strains in Figure 11.6 are presented in Table 11.2 and Figure 11.7. Good agreement is obtained except for cycle 5. It is also seen that the R values are reasonably invariant with respect to crack length.

Table 11.2

Neutral axis location

Cycle	P kN	a mm	K_I kN-m$^{-3/2}$	R Theo. Eq. 8	R Exper.	Theo./Exp.
1	1.11	41.1 (a_o)	337	0.422	0.453	0.93
2	1.07	55.9	486	0.426	0.389	1.10
3	1.11	64.0	653	0.437	0.426	1.03
4	1.11	69.6	798	0.448	0.508	0.88
5	0.65	79.8	729	0.475	0.337	1.41
			avg.	0.442	0.432	

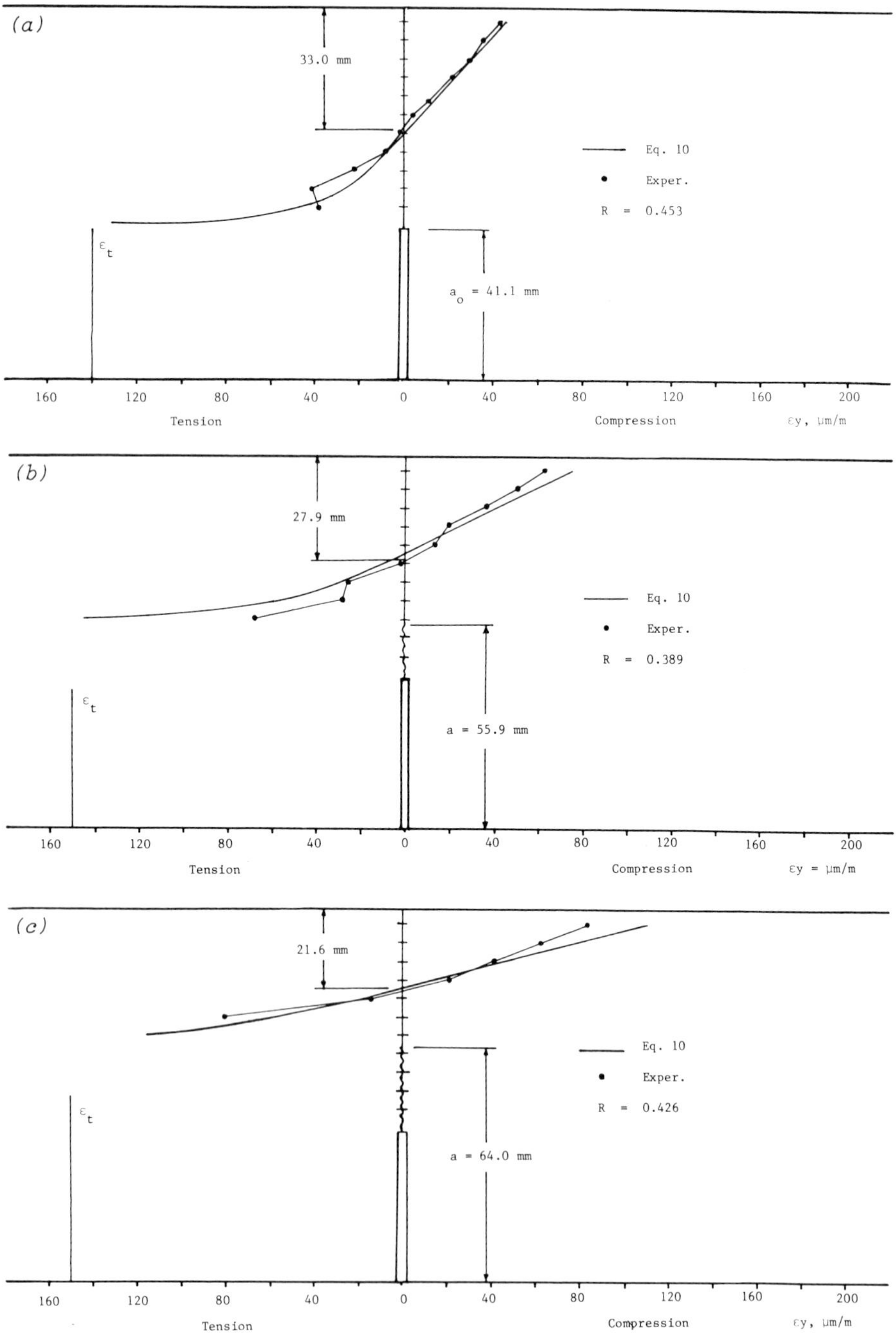

Figure 11.6: *(a) Strain distribution,* $\frac{a_o}{W} = 0.40$, *P = 1112 N*
(b) Strain distribution, $\frac{a}{W} = 0.55$, *P = 1068 N*
(c) Strain distribution, $\frac{a}{W} = 0.63$, *P = 1112 N*

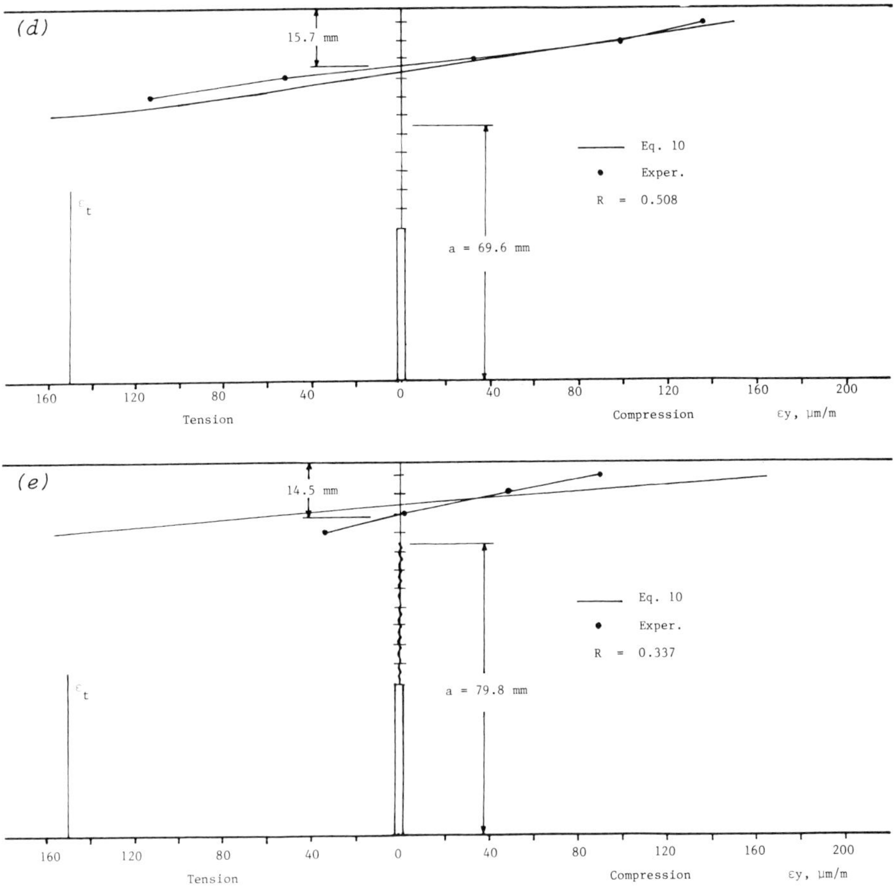

Figure 11.6: (d) Strain distribution, $\frac{a}{W} = 0.68$, *P = 1112 N*

(e) Strain distribution, $\frac{a}{W} = 0.78$, *P = 645 N*

Critical crack-tip-opening-displacement

In order to use Equation (11.1) to determine CTOD it was necessary to determine an average value of η. This assumes that regardless of crack length, the shape of the crack front is the same. Using the results of dye tests on twenty-eight beams the average value of η was determined to be $\eta = 5.4$mm. The average, theoretical value of R is 0.442 (Table 11.2) and the knife edge thickness z is 3.2mm. Then, Equation (11.1) becomes

$$\delta_c = \frac{0.442\ (W - a) + 5.4}{0.442\ (W - a) + a + 3.2}\ v$$

or

$$\delta_c = \frac{50.3 - 0.442\ a}{48.1 + 0.558\ a}\ v \qquad (11.11)$$

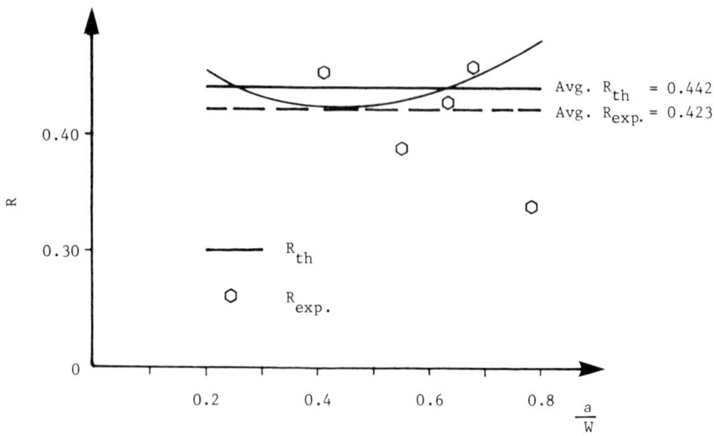

Figure 11.7: Variation of R with crack depth

In order to determine v, the CMOD associated with $CTOD_c$, the following argument is used. The crack-tip-opening displacement δ is zero (except for a negligible elastic effect) until the region in front of the crack tip starts to experience subcritical crack growth. This occurs at the onset of non-linearity in the plot of P versus CMOD (cf, Figure 11.5). This point is approximately at P = 0.7 P_{MAX}.

The subcritical crack growth continues until the process zone is fully developed and unstable crack growth occurs. This takes place approximately at P = 0.95 P_{MAX} on the softening (descending) part of the curve. At this point δ is assumed to reach the critical value δ_c. The value of v used in Equation (11.11) then is obtained from

$$v = CMOD\Big|_{.95\ P_{MAX}} - CMOD\Big|_{.70\ P_{MAX}} \tag{11.12}$$

Values of P_{MAX}, and v for each load cycle, obtained directly from Figure 11.5 are presented in Table 11.3 along with computed values of δ_c. Considering all values, the average δ_c is 11.1 µm + 50%, - 33%.

CONCLUSIONS

The measured strain values in the region in front of the crack agree reasonably well with those predicted despite the problems described earlier. The results for location of the neutral axis obtained experimentally also agree well with theoretical predictions and are approximately the same for different crack lengths.

Reasonable estimates for the critical crack-tip-opening-displacement for beams in bending thus appear to be possible with the simplified model presented which is based on LEFM. Whether these estimates can be correlated with $CTOD_c$ values obtained from tensile specimens remains a subject for further research.

Table 11.3

Critical crack-tip-opening displacement

Cycle	a mm	P_{MAX} kN	v µm	δ_c µm
1	41.4 (a)	2.94	44.5	16.7
2	55.9	2.60	43.2	11.0
3	64.0	1.92	55.9	11.1
4	69.6	1.25	57.2	9.3
5	79.8	0.74	71.1	7.4

ACKNOWLEDGEMENTS

The research reported herein was supported by the Engineering Experiment Station, Kansas State University and by the National Science Foundation through grants CEE-8305120, CEE-8317136 and MSM-8317136. This support is gratefully acknowledged.

REFERENCES

[1] Swartz, S.E. and Refai, T.M.E., "Influence of Size Effects on Opening Mode Fracture Parameters for Precracked Concrete Beams in Bending", Proceedings International Conference on Fracture of Concrete and Rock, Houston (June 1987)

[2] Wecharatana, M. and Shah, S.P., "A Model for Predicting Fracture Resistance of Fiber Reinforced Concrete", Cement and Concrete Research, 13 (1983)

[3] Wecharatana, M. and Chiou, W.J., "Fracture Energy and the Complete Tensile Softening Response of Concrete", Proceedings of the 1986 SEM Spring Conference on Experimental Mechanics, New Orleans (June 1986)

[4] Jenq, Y.S. and Shah, S.P., "Two Parameter Fracture Models for Concrete", Journal of Engineering Mechanics, ASCE, 111, No. 10 (1985)

[5] Rood, S., "Fracture Toughness Testing of Small Concrete Beams" MS Thesis, Civil Eng. Dept., Kansas State Univ., Manhattan, Kansas, 1984

[6] Go., C.G. and Swartz, S.E., "The COD as a Fracture Criterion for Concrete", Proceedings 5th ASCE-EMD Speciality Conference, University of Wyoming (August 1984)

[7] Knott, J.F., "Fundamentals of Fracture Mechanics", Halsted Press (1973)

[8] Go., C.G., Swartz, S.E., and Hu, K.K., "Stress Intensity Factors for Single-Edge-Notch Beam", Journal of Engineering Mechanics, ASCE, 110, No. 4 (1984)

(9) Swartz, S.E., Hu K.K. and Jones, G.L., Techniques to Monitor Crack Growth in Plain Concrete Beams," Experimental Techniques, 6, No. 6, (1982).

(10)Cedolin, L., DeiPoli, S. and Iori, I., Tensile Behavior of Concrete, "Journal of Engineering Mechanics, ASCE, 113, No. 3, (1987).

(11)Swartz, S.E. and Go, C. G., "Validity of Compliance Calibration to Cracked Concrete Beams in Bending," Experimental Mechanics, 24, No 2, (1984).

12 Measurement of boundary effects in surface flaws

C.W.Smith & M.Rezvani

SUMMARY

The present study involves the use of improved algorithms for converting optical data from frozen stress and high density Moiré interferometry into appropriate fracture parameters for evaluating the influence of a free surface intersected by a crack on the order of the stress singularity at the crack tip in nearly incompressible bodies. Frozen stress and Moiré interferometric data obtained from stress frozen compact bending specimens were used to develop algorithms for converting frozen stress data into stress intensity factors which account for the free surface effect. Using the above mentioned algorithms, these data were used to obtain the singularity order distribution along surface flaw borders. Finally, stress intensity factor distributions neglecting and accounting for the boundary effect are computed and compared to demonstrate the free surface effect on the stress intensity distributions.

NOTATION

r,θ	polar coordinates with origin at crack tip (Figure 12.1a)
n,z,t	local rectangular cartesian coordinates at crack tip (Figure 12.1b)
a	crack length
σ_{ij}	stress components;
σ^0_{ij}	non-singular stress components;
$\sigma^0/\sqrt{8}$	non-singular contribution to τ_{max}.
τ_{max}	maximum shear stress in x_1-x_2 or nz planes;
τ_0	contribution to τ_{max} of non-singular stresses
K	Mode I Stress Intensity Factor (SIF);
K_λ	stress eigenfactor;
K_{cor}	corresponding SIF

(K_{Ap}), $(K_\lambda)_{Ap}$, $(K_{cor})_{Ap}$	apparent values defined in text
$f_{ij}(\theta)$, $G_i(\theta)$, $F_{ij}(\theta)$, $D_i(\theta)$	functions of θ
C, D_z	coefficients containing elastic constants
λ_σ, λ_u	lowest dominant eigenvalues near free surface intersected by a crack at right angles in stress; displacement field equations

INTRODUCTION

The apparent loss in the inverse square root stress singularity when a crack intersects a free surface has been an issue of some controversy since around 1970 [1]. However, in 1980, Benthem [2,3] provided an analytical solution for a quarter infinite crack intersecting the free surface of a half space. Subsequently, this result was confirmed by finite element analyses [4,5] and by experiments conducted on compact-bending specimens [6,7]. The effect was found to be sufficiently significant in incompressible materials to affect the accuracy of surface measurements on them. Moreover, the measurements suggested the presence of a transition zone inside the body, over which the singularity order varied from its surface value to the classical value of one half. This was in contrast to assumptions and predictions from analytical solutions.

For these reasons, subsequent experiments were conducted in order to evaluate the effect of more practical flaw geometries, namely semi-elliptic surface flaws [8,9]. These studies showed enough variation in the results to suggest that improved algorithms would be desirable for such problems.

The experimental methods used in the experiments described above [8] involved the tandem application of frozen stress photoelasticity and high density Moiré interferometry. The approach was based on studies by the first author and his colleagues which spans nearly two decades and which was directed towards the development of cost effective optical methods for evaluating three dimensional effects in cracked body problems. In general, the full method of analysis involves the following steps:

(i) Construct a scale model of the body from suitable stress freezing material and insert a natural or artificial crack. Natural cracks are produced by striking a sharp blade held normal to the specimen surface with a hammer. The starter crack emanates from the blade tip, propagates into the body and then arrests. Further crack growth is produced in the stress freezing oven under load while observed through an oven port. When the desired crack size is reached, the load is reduced to terminate flaw growth. "Artificial" cracks are simply vee notches machined into the model and are not extended under load. Extensive studies by Smith et al [10] suggest that the stress fringe patterns will be virtually identical to those for cracks outside a small near tip non-linear zone provided the included notch angle is not greater than 30°.

(ii) Freeze in the stress fringe and deformation fields produced above critical temperature under load.

(iii) Remove thin slices along the crack border which are mutually orthogonal to the crack surface and the crack border.

(iv) Analyze the frozen slices photoelastically to obtain fringe field distributions using combined Tardy Compensation and partial mirror fringe multiplication and a digitizer.

(v) Apply linear gratings to a free surface of the "frozen" slices.

(vi) Anneal slices to deform gratings.

(vii) Observe deformed grating pattern through a "virtual" (optical) master grating to obtain Moiré displacement fringe distributions.

(viii) Using appropriate algorithms, extract appropriate fracture parameters from the results of steps (iv) and (vii) and compare results.

The goal of the present chapter is to validate an improved frozen stress algorithm by comparing results from its use on a compact bending specimen with previously obtained Moiré data and then to use the frozen stress method to analyze the free surface effect for natural surface flaws under Mode I loading. Finally, a means for interpreting results within the context of Linear Elastic Fracture Mechanics (LEFM) is suggested.

ANALYTICAL CONSIDERATIONS

One may begin by reviewing the two parameter approach which has been employed extensively [11,12,13] in LEFM.

LEFM frozen stress algorithm - two parameter approach

By choosing a data zone sufficiently close to the crack tip that a Taylor Series Expansion of the non-singular stresses can be truncated to the leading terms, one may write [11] (see Figure 12.1a)

$$\sigma_{ij} = \frac{K}{(2\pi r)^{1/2}} f_{ij}(\theta) - \sigma^0_{ij}(\theta) \qquad i,j = 1,2 \tag{12.1}$$

where σ_{ij} are the stress componments, K is the Mode I Stress Intensity Factor (SIF) and r, are polar coordinates with origin at the crack tip. Along $\theta = \pi/2$, one has, for case where $(\sigma^0)^2$ is small relative to $8\tau_m^2$ [11] (where σ^0 is proportional to the contribution of the non-singular stress to the maximum shear stress, τ_{max})

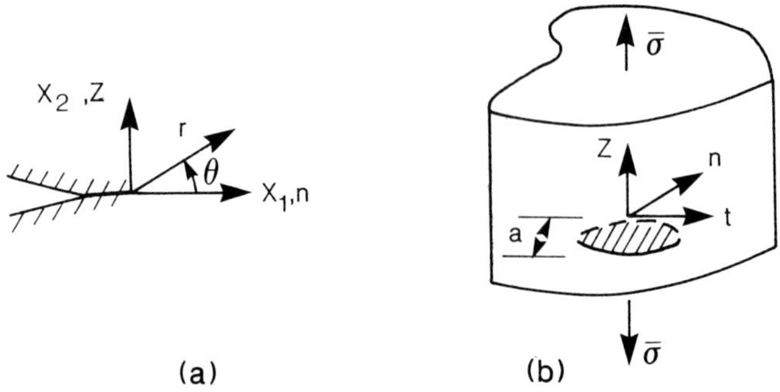

Figure 12.1

$$\tau_{max} = \frac{K_I}{(8\pi r)^{1/2}} + \frac{\sigma^0}{\sqrt{8}} \tag{12.2}$$

Now defining an "apparent" SIF,

$$K_{Ap} = \tau_{max}(8\pi r)^{1/2} \tag{12.3}$$

and normalizing with respect to $\bar{\sigma}(\pi a)^{1/2}$ where $\bar{\sigma}$ represents the remote stress and $\underline{a}$ the crack length, one has

$$\frac{K_{Ap}}{\bar{\sigma}(\pi a)^{1/2}} = \frac{K}{\bar{\sigma}(\pi a)^{1/2}} + \frac{\sigma^0}{\bar{\sigma}}\left(\frac{r}{a}\right)^{1/2} \tag{12.4}$$

which suggests a linear zone in a plot of $\frac{K_{Ap}}{\bar{\sigma}(\pi a)^{1/2}}$ vs $\left(\frac{r}{a}\right)^{1/2}$ with a slope of $\frac{\sigma^0}{\bar{\sigma}}$. Experience shows this zone to lie between $(r/a)^{1/2}$ values of approximately 0.2 to 0.4 (or above) in two dimensional problems. By extracting optical data from this zone and extrapolating across a near tip non-linear zone, an accurate estimate of $K_1/\bar{\sigma}(\pi a)^{1/2}$ can be obtained (Figure 12.2). This approach may be extended directly to three dimensional problems by replacing i,j values of 1,2 by n,z where the nz plane is normal to the crack border at each point along the flaw border and the nzt coordinate system is a local cartesian system which follows the crack border (Figure 12.1b). In three dimensional problems the outer boundary of the data zone is usually restricted to $(r/a)^{1/2} = 0.4$ or less.

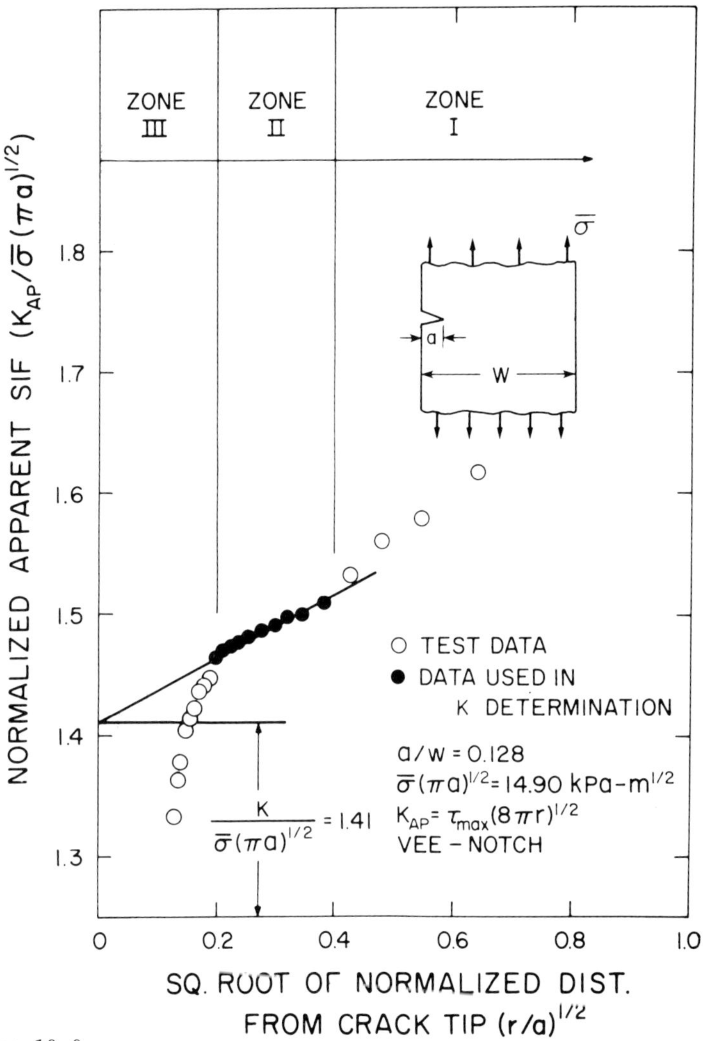

Figure 12.2

LEFM moiré algorithm - two parameter approach

Corresponding to Equation (12.1), one has for displacements for the two dimensional case [14]

$$u_i = C\,K\,G_i(\theta)r^{1/2} \begin{cases} -\dfrac{\sigma^0}{E_s} r\cos\theta & i = 1 \\ +\dfrac{\sigma^0}{E_s} r\sin\theta & i = 2 \end{cases} \tag{12.5}$$

where C is a constant containing the elastic constants, σ^0 is a non-singular stress term and E_s is the shear modulus.

In the frozen stress algorithm, the effect of the non-singular stress was included since it was independent of r. However, in Equation (12.5) one may, for sufficiently small values or r, neglect the effect of the σ^0 terms. Thus, along $\theta = \pi/2$, for a three dimensional problem, one may write

$$u_z = C'\,K\,r^{1/2} \tag{12.6}$$

where a plot of $u_z/r^{1/2}$ vs. $r^{1/2}$ should yield a straight horizontal line whose ordinate is K. If one is not sufficiently close to the crack tip, however, the σ^0 terms do make some contribution in the data zone so one usually uses the form

$$u_z = C'\,K_{Ap}\,r^{1/2} \tag{12.7}$$

and extrapolates the linear part of u_z vs. $r^{1/2}$ to the origin as before.

Algorithms accounting for free boundary effect

As noted earlier, when a crack intersects a free surface at right angles, the dominant eigenvalue is altered. Using Benthem's three dimensional variables, separable eigenfunction expansion of the σ_{ij} and u_i near the crack tip at the free surface for a quarter infinite crack intersecting a half space at right angles and the LEFM results as a guide, one can construct the following functional forms for the near tip σ_{ij} and u_i:

$$\sigma_{ij} = \frac{F_{ij}(\lambda_\sigma, \theta)}{r^{\lambda_\sigma}} - \sigma^0_{ij} \qquad (i,j = n,z) \tag{12.8}$$

where $\sigma^0_{ij} = \sigma^0$ for i,j = n and zero for all other values of i,j.

$$u_i = D_i(\theta)r^{\lambda_u} \tag{12.9}$$

where λ_σ and λ_u represent the lowest dominant eigenvalues in the local stress and displacement field equations, respectively. Here similar functional forms are implied near the crack tip for σ_{ij} and u_i as in LEFM but λ is undetermined. From Equation (12.9), one may, along $\theta = \pi/2$ write

$$u_z = D_z r^{\lambda_u}$$
$$\log u_z = \log D_z + \lambda_u \log r \qquad (12.10)$$

and determine λ_u from a plot of $\log u_z$ vs. $\log r$ in the linear range. An example of this procedure is given in Figure 12.3 using Moiré data [15].

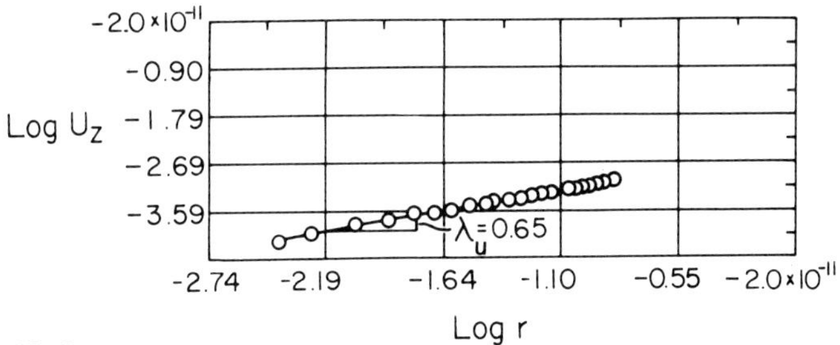

Figure 12.3

Independent determination of λ_σ may also be carried out but the procedure is not as straightforward due to the presence of the second term in Equation (12.8) and the highly three dimensional nature of the near tip stress state at the free surface. If τ_{max} is evaluated from Equation (12.8) along $\theta = \pi/2$ as done in LEFM, one obtains:

$$\tau_{max} = \frac{\lambda_\sigma K_{\lambda_\sigma}}{\sqrt{2\pi}\, r^{\lambda_\sigma}} + \frac{\sigma^0}{2}\sin(\lambda_\sigma+1)\frac{\pi}{2} = \lambda_\sigma \frac{(K_{\lambda_\sigma})_{Ap}}{\sqrt{2\pi}\, r^{\lambda_\sigma}} \qquad (12.11)$$

where K_{λ_σ} may be designated a stress eigenfactor. Defining $\tau_0 = \frac{\sigma^0}{2}\sin(\lambda_\sigma+1)\frac{\pi}{2}$, one obtains

$$\log(\tau_{max}-\tau_0) = \log\left\{\frac{\lambda_\sigma K_{\lambda_\sigma}}{\sqrt{2\pi}}\right\} - \lambda_\sigma(\log r) \qquad (12.12)$$

Thus, if τ_0 can be determined, one can plot $\log(\tau_{max} - \tau_0)$ vs. $\log r$ and obtain λ_σ as the slope of the linear range. Since LEFM is expected to prevail away from the free surface, one may conjecture that the value of τ_0 as determined from LEFM away from the free surface should suffice as an adequate correction for the non-singular field effect except near the free surface.

Equation (12.2) shows that in LEFM, $\tau_0 = \frac{\sigma^0}{\sqrt{8}}$. Thus τ_0 can be determined from the slope of the normalized K_{Ap} vs. $r^{1/2}$ graph in the interior of the body (see Equation (12.4). Near the free surface, however, since λ_σ is unknown, τ_0 cannot be determined in the same way. Instead the result obtained from Moire analysis of the free surface is relied upon.

EXPERIMENTAL RESULTS

In the present study, experiments on compact bending specimens (Figure 12.4) were first conducted using both frozen stress and Moiré interferometric methods. Frozen stress data were analyzed in the interior with Equation (12.12) with τ_0 being obtained from LEFM. Results together with Moiré results obtained near the free surface from a plot of Equation (12.10) and Benthem's result are all shown in Figure 12.5a. These results confirm that the LEFM evaluation of τ_0 appears adequate in the interior of the body. However the best correlation between the photoelastic and other results near the free surface was obtained by decreasing τ_0 from the LEFM value to zero at the free surface linearly from the onset of the transition zone in λ_σ for analysis of the photoelastic data from slices near the surface, and then using the surface slice result as a free surface value. Due to the unavoidable data scatter, it would not appear that more accurate evaluations of τ_0 are warranted here. Thus, the use of Moiré data at the free surface is circumvented in this approach once it has been used to validate the assumption that $\tau_0 = 0$ near the free surface.

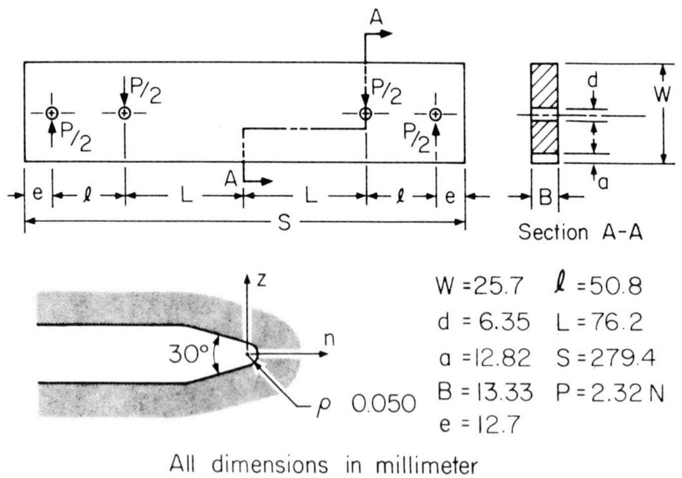

Figure 12.4

Having "benchmarked" the frozen stress method in the compact bending experiments against both Bentham's and Moiré results, the frozen stress method was next applied to a study of surface flaws in wide plates under Mode I loading. The distributions of λ_σ for all tests are given in Figure 12.5b, with τ_0 determined

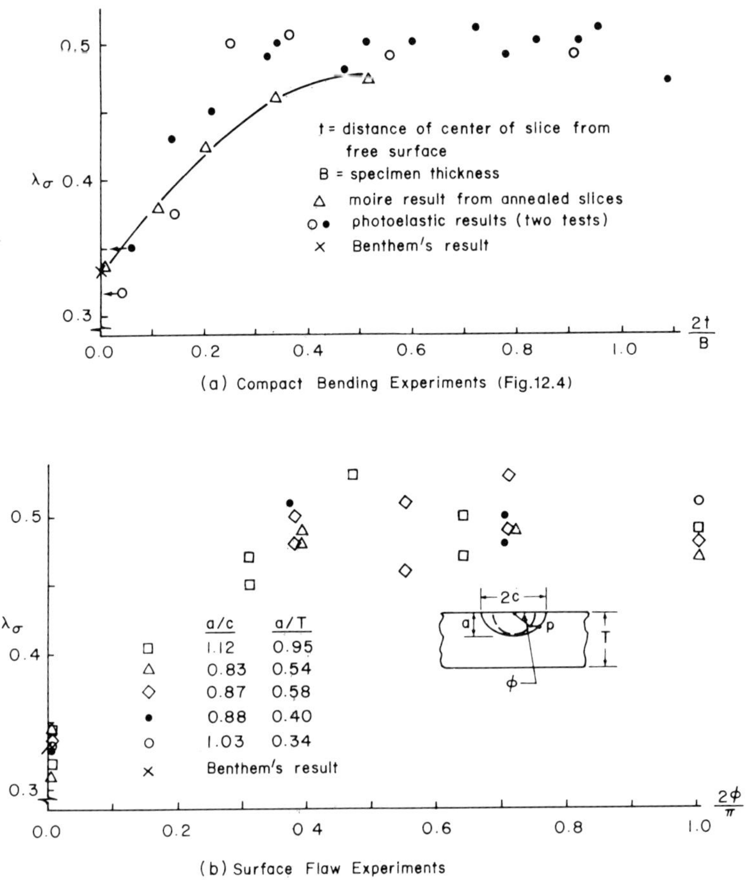

Figure 12.5: *(a) Compact bending experiments (Figure 12.4)*
(b) Surface flaw experiments

in the same way as for the compact-bending specimens. These results suggest that the effects of changes in the crack shape and size of the surface flaws are only marginally larger than the scatter between replicated photoelastic tests on straight front cracks as shown in Figure 12.5a and the Moiré results on one of the same specimens near the free surface. Data zones for the photoelastic determination of λ_σ were all between values of $(r/a)^{1/2}$ = 0.25 and 0.45 and for the Moiré data somewhat further from the crack tip due to lack of complete reversibility in the annealing process. (While reversibility of annealing is not complete and pushes the data zone further from the crack tip, it has only a negligible effect on the SIF in LEFM [16]).

USE OF RESULTS IN FRACTURE MECHANICS

Since the value of the lowest dominant eigenvalue differs from that associated with classical fracture mechanics, one is led to address the issue of how the quantified knowledge of this phenomena can be introduced into the framework of classical fracture mechanics so it can be utilized in engineering analysis. A simple way to achieve this is to introduce a stress intensity factor which would be required to produce the same value of τ_{max} as one measures in the actual test where λ_σ is not equal to the classical value. Calling this the corresponding SIF,

$$\frac{(K_{cor})_{Ap}}{r^{1/2}} = \frac{(K_{\lambda_\sigma})_{Ap}}{r^{\lambda_\sigma}} \qquad (12.13)$$

Now, substituting relation (12.13) into Equation (12.11), one has:

$$\tau_{max} = \frac{\lambda_\sigma (K_{cor})_{Ap}}{\sqrt{2\pi}\ r^{1/2}} \qquad (12.14)$$

Now, combining Equations (12.3) and (12.4), we have,

$$\frac{\lambda_\sigma (K_{cor})_{Ap}}{\sqrt{2\pi}\ r^{1/2}} = \frac{K_{Ap}}{\sqrt{8\pi}\ r^{1/2}} \quad \text{whence}$$

$$(K_{cor})_{Ap} = \frac{K_{Ap}}{2\lambda_\sigma} \qquad (12.15)$$

Since values of K and K_{cor} are obtained by extrapolating plots of K_{Ap} vs. $r^{1/2}$ across a non-linear zone to the origin, it follows

$$K_{cor} = \frac{K}{2\lambda_\sigma} \qquad (12.16)$$

It should be emphasized that the introduction of relation (12.13) is quite arbitrary and is based upon purely dimensional grounds. Moreover, since the analysis here is quasi-two dimensional, one depends on Moiré results using Benthem's analysis to validate our stress algorithm at the free surface.

As can be seen in Figure 12.6, the use of K_{cor} near the free surface tends to raise the K distribution which would be obtained from LEFM. The K distributions were obtained using the LEFM algorithms on the experimental optical data and were in general agreement with LEFM numerical analyses. The difference between K_{cor} and K away from the free surface in the surface flaw (Figure 12.6b) is conjectured to be related to the error in assuming that $\lambda_\sigma = 1/2$ in the middle portion of the surface flaw border when computing τ_0.

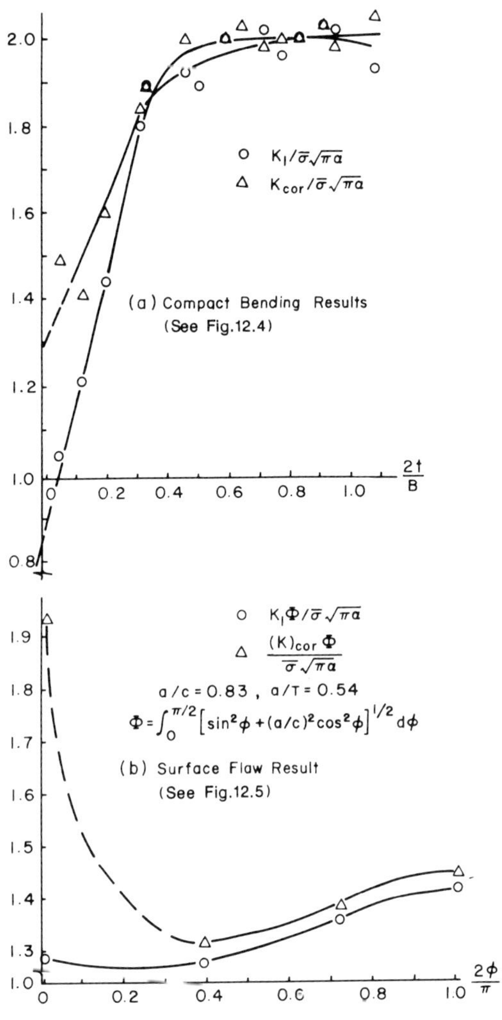

Figure 12.6: (a) *Compact bending results (see Figure 12.4)*
(b) *Surface flaw result (see Figure 12.5)*

These results suggest that, when measurements are made on or near the surface intersected by a crack in nearly incompressible materials and interpreted within the framework of LEFM, the SIF will be nonconservative and should be modified to account for the free surface effect.

CONCLUSIONS

Algorithms and results have been presented for using frozen stress photoelastic and Moiré measurements to determine the influence of a free surface intersected at right angles by a crack upon the order of the dominant near tip eigenvalue. The methods were first validated using compact-bending specimens with straight-front through cracks and then applied to surface flaws. Finally a method was suggested for accounting for the measured effects within the context of LEFM.

Test results revealed data scatter and algorithms were only approximate. However, results are believed to be within reasonable engineering accuracy (i.e. ± 5%) and the use of the corresponding SIF appears to be a simple way of accounting for the measured nonconservative effect in measurements on nearly incompressible materials.

ACKNOWLEDGEMENTS

The authors wish to acknowledge the contributions of Prof. J.P. Bentham, Prof. D. Post, Dr. J.S. Epstein, Dr. O. Olaosebikan, W.R. Lloyd and T.J. Theiss to the success of the present studies. They also gratefully acknowledge the support of the National Science Foundation under Grant No. MEA-832-0252.

REFERENCES

[1] Sih, G.C. "A Review of the Three-Dimensional Stress Problem for a Cracked Plate", Intl. Journal of Fracture Mechanics, Vol. 7, No. 1, March 1971, pp. 39-61

[2] Bentham, J.P., "On an Inversion Theorem for Conical Regions in Elasticity Theory", J. of Elasticity, Vol. 9, No. 2, 1979, pp. 159-169

[3] Bentham, J.P., "The Quarter Infinite Crack in a Half-Space; Alternative and Additional Solutions", Intl. J. of Solids & Structures, Vol. 16, 1980, pp. 119-130

[4] Bazant, J.P. and Estensoro, L.F., "Stress Singularity and Propagation of Cracks at Their Intersection with Surfaces", Intl. J. of Solids & Structures, Vol. 16, 1980, pp. 119-130

[5] Burton, W.S., Sinclair, G.B., Solecki, J.S. and Swedlow, J.L.,m "On the Implications for LEFM of the Three Dimensional Aspects in Some Crack/Surface Intersection Problems", Intl. J. of Fracture, Vol. 25, 1984, pp. 3-32

[6] Smith, C.W. and Epstein, J.S., "Measurement of Three Dimensional Effects in Cracked Bodies", Proc. of Vth Intl. Congress on Experimental Mechanics, 1984, pp. 102-107

[7] Smith, C.W. and Epstein, J.S., "Boundary Layer Measurements in Cracked Body Problems", Proc. of Twelfth Southeastern Conf. on Theoretical & Applied Mechanics, Vol. 1, 1984, pp. 355-360

[8] Smith, C.W., Olaosebikan, O. and Epstein, J.S., "Boundary Effects in Fracture Mechanics", Proc. of Annual Conf. of the Society for Experimental Mechanics, 1985, pp. 113-118

[9] Ruiz, C. and Epstein, J.S., "On the Variation of the Stress Intensity Factor Along the Front of a Surface Flaw", Intl. J. of Fracture, Vol. 28, 1985, pp. 231-238

[10] Smith, C.W., McGowan, J.J. and Peters, W.H., "Effects of Artificial Cracks and Poisson's Ratio Upon Photoelastic Stress Intensity Determination", J. of Experimental Mechanics, Vol. 16, No. 5, May 1976, pp. 188-193

[11] Smith, C.W., "Use of Three Dimensional Photoelasticity and Progress in Related Areas", Ch. 1 of Experimental Techniques in Fracture Mechanics, Vol. 2, A.S. Kobayashi, Ed., Society for Experimental Stress Analysis, 1975, pp. 3-58

[12] Smith, C.W., "Use of Photoelasticity in Fracture Mechanics", Ch. 2 of Experimental Evaluation of Stress Concentration and Intensity Factors, Mechanics of Fracture, Vol. 7, G.C. Sih, Ed., Martinus-Nijhoff Publishers, 1981, pp. 163-188

[13] Smith, C.W. and Kobayashi, A.S., "Experimental Fracture Mechanics", Ch. 2 of Handbook on Experimental Mechanics, Society for Experimental Mechanics, Prentice-Hall, 1987, pp. 891-956

[14] Eftis, J., Subramonian, N. and Liebowitz, H., "Crack Border Stress and Displacement Equations Revisited", J. of Engr. Fracture Mechanics, Vol. 19, No. 1, 1977, pp. 189-210

[15] Smith, C.W. and Epstein, J.S., "Measurement of Three Dimensional Effects in Cracked Bodies", Proc. of Vth Intl. Congress on Experimental Mechanics, June 1984, pp. 102-110

[16] Smith, C.W., Post, D. and Nicoletto, "Experimental Stress Intensity Distributions in Three Dimensional Cracked Body Problems", Proc. of a Joint Conference on Experimental Mechanics, Society for Experimental Stress Analysis and Japan Society of Mechanical Engineers, Part I, 1982, pp. 196-200

13 Developments in measuring techniques for high-rate dynamic fracture testing

H.J.MacGillivray, E.R.Akum, P.Brooks-Johnson & V.Grabulov

SUMMARY

Measurement of force, displacement and crack behaviour are fundamental to the understanding of dynamic fracture in engineering materials. The most widely used methods now available are described and illustrated by some results of current research programmes.

INTRODUCTION

Fractures originating from cracks and small defects are a particular problem in heavily stressed engineering components and structures, as such defects are almost unavoidable. Fracture mechanics attempts to predict the critical size of crack or severity of loading to cause failure, which requires a knowledge of fracture related material properties.

Testing techniques for static loading conditions are now well developed and a number of standard methods are in general use [1-4]. Many engineering components and structures are however subject to dynamic loads in normal use or in accidental overload conditions. Applications where dynamic loading must be considered include all the transportation industries, heavy machinery, nuclear plant, military vehicles and any equipment subject to impact, explosive or seismic loading. The materials involved include structural and alloy steels, aluminium, titanium, polymers, composites and ceramics. Loading rates range from quasi-static to about $50ms^{-1}$ for conventional testing, but much higher for ballistic impact studies. The data required are normally obtained in laboratory scale tests, using standard or near-standard test pieces, to allow comparison with static results.

One serious problem is the comparison of results obtained in different laboratories using often widely varying techniques. Some progress has been made towards standardisation for measurement and calibration with the publication of the new British Standard BS 6729:1987 [5] and with Annex A7 to the ASTM Standard E399 [1] but at present both methods are limited to the lower end of the loading rate range, about $1ms^{-1}$.

The most important measurements which need to be made during fracture testing are force, displacement and initiation and propagation of cracks which may involve strain gauge, electrical or optical methods. Results must be obtained under difficult conditions on a fast-moving test piece, and the effects of inertia and stress-waves must be evaluated. The cost of manufacturing and preparing each specimen can be high, so maximum reliability of measurement and recording systems is essential. Each measuring method is examined in detail comparing such factors as accuracy and repeatability, reliability, ease of use, cost and availability of the most generally used techniques. The results of some current research programmes at Imperial College are presented as examples of the range of dynamic fracture testing and the importance of correct measurement.

THE DEVELOPMENT OF DYNAMIC FRACTURE TESTING

The simplest forms of dynamic test use the Charpy and Izod pendulum impact machines developed in the early years of this century. A small blunt-notched specimen is used and only the total energy absorbed in fracture can be measured. This limits the method mainly to quality control for manufacturing or an aid to materials development in conjunction with static fracture toughness testing. The addition of a force measuring transducer to the striker, first made by Tanaka and Umekawa [6] in the late 1950's, allowed the load on the specimen to be measured reliably during the impact event. Instrumented Charpy machines with improved transducers, recorders and data analysis systems and with capacities up to 600 joules are now widely used in testing laboratories. Large pendulum and dropweight impact machines with capacities up to 0.5 Mjoules are supplemented by high-speed servohydraulic testing machines. All these provide impact velocities in the range 5-20ms^{-1}. For higher rates, gas operated projectile machines are used often combined with the Hopkinson bar principle where pressure waves can provide pulses of varying amplitude and duration at velocities up to 50ms^{-1}.

The measurement problems encountered in dynamic fracture testing are illustrated by the composite diagram Figure 13.1. This shows a typical unfiltered load-time record for the ductile fracture of a pre-cracked Charpy vee-notch (PCVN) 3-point bend specimen with superimposed lines indicating fully brittle, post-yield brittle and ductile growth before cleavage fracture modes. This type of test was first reported by Radon and Turner [7] in the 1960's and is now widely used particularly for K_{1D} measurement in pressure vessel steels by, for example, Server and Tetelman [8] and more recently Kobayashi [9]. The record is composed of a decaying oscillating waveform which rises to a maximum then slowly decreases. Because of the stress concentration effect, stress levels exceeding yield are reached at the crack tip long before general yielding occurs. A material with low toughness, or a tough material tested at low temperature, may fail by cleavage in the linear region, otherwise local yielding called the

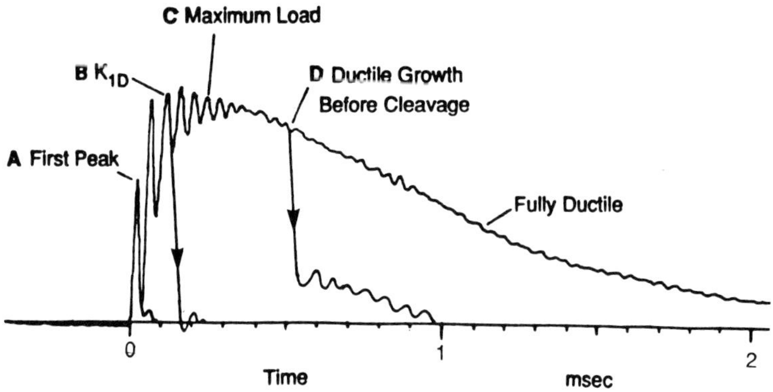

Figure 13.1: Dynamic fracture types

crack tip plastic zone, develops up to the maximum load and the crack tip blunts. At some point during this process the crack initiates and begins to extend by ductile tearing, which continues until complete failure occurs by tearing or cleavage.

For fracture in the linear range or where only a small amount of crack tip plasticity is present, linear elastic fracture mechanics (LEFM) concepts can be used. Here the basic relationship for Mode 1 loading is:

$$K = Y \sigma a^{0.5} \qquad (13.1)$$

where
Y is a constant
σ is the remote stress
a is the crack length

and the critical value of the crack tip stress intensity factor K at fracture under plane strain conditions which impose the greatest constraint is the plane strain fracture toughness K_{1C} often denoted K_{1D} if the loading is dynamic. Thus to determine K under dynamic conditions, measurement of force is necessary together with some reliable indication of crack initiation and extension.

The magnitude of the first load oscillation depends primarily on the inertia of the specimen rather than its toughness, so very brittle fractures occuring at point A on the load-time record in Figure 13.1 cannot be analyzed directly. Either an inertia correction is required [8], or the load-time trace can be used to fit a computed solution with a dynamic finite element programme as in Curr et al [10] and Kobayashi [11]. The alternative is to reduce the striker velocity, the 'low-blow' test, which decreases the inertia effect but lowers the loading rate.

When fracture occurs after two or three oscillations of the load signal, as shown at point B in Figure 13.1, the dynamic toughness K_{1D} can be calculated using LEFM. The requirements of the test record and calibration of the equipment are covered by the new British Standard BS 6729 [5] and by the standard proposed but not yet adopted by ASTM committee E 24 [12].

Significant crack tip yielding occurs beyond point B, invalidating LEFM laws, and elastic-plastic fracture mechanics (EPFM) must be used. The crack tip opening displacement (CTOD) technique adapted for dynamic fracture is fully described in [5] and requires measurement of force and either crack opening or load point displacement.

The other well developed method is the J contour integral, measuring the rate at which elastic strain energy is being dissipated at the crack tip and usually evaluated by a relationship of the form:

$$J = \eta U / B(W-a) \tag{13.2}$$

J and CTOD can be related by the expression:

$$J = M \sigma_y \delta \tag{13.3}$$

where
η is a constant which may very for dynamic loading
U is the absorbed energy
B is the specimen thickness
W is the specimen width
M is the stress intensification factor
σ_y is the yield stress
δ is the CTOD

In experimental studies the value of the term U is found from the area under the load-displacement curve, and the results expressed as a plot of J against crack length or J-R curve which is extrapolated, either to zero growth or to a small amount of growth determined by blunting, to find initiation toughness. The dynamic force measurement is again required plus load point displacement, while the initiation of ductile crack growth cannot be detected directly on the load record.

For comparison purposes maximum load toughness can be assumed, point C, in Figure 13.1, although as McConnell and Server [13] have stated this may be non-conservative. Alternatively, over a small temperature band in the transition temperature range of low alloy steels, a small amount of ductile crack growth can occur before cleavage at point D. In this particular case if the temperature differences are neglected a limited J-R curve can be obtained by plotting J up to the cleavage point against the ductile growth measured on the fracture surface [10] (Figure 13.2). The difficulty of detecting initiation from the load-time record alone has led to the development of a large number of test techniques for the independent measurement of crack initiation.

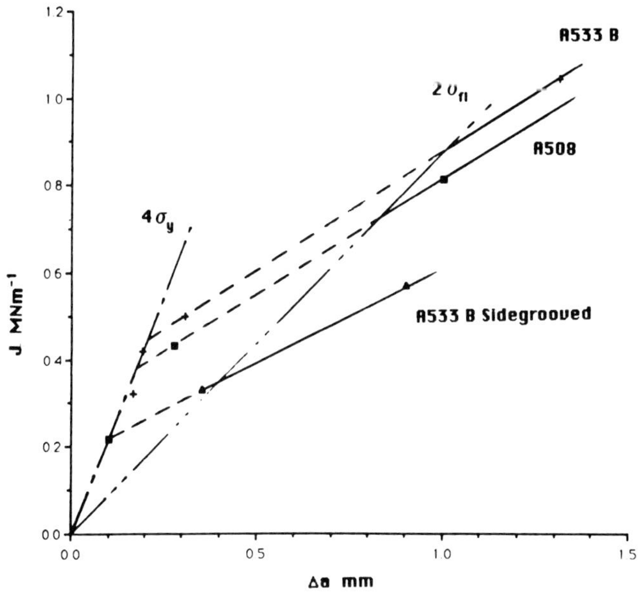

Figure 13.2: Cleavage R-curve

MEASUREMENTS IN DYNAMIC FRACTURE

The principal measurements made in dynamic fracture testing are first the force applied to the specimen by the movement of the striker. Displacement of the load point is needed for impact velocity. CTOD and energy measurement, while displacements actually on the testpiece can be used to detect crack initiation, directly measure CTOD and infer crack tip K values. Many techniques are available for detecting crack initiation independently, and finally the length of the growing crack is required for crack velocity studies. As some techniques can be applied to measure more than one parameter - foil strain gauges can measure all four - each technique will be described and its applications considered.

Considerations which apply to all techniques are calibration and frequency response. The conditions for testing transducers, amplifiers and recording equipment are discussed in detail in Appendix B of the new British Standard [5] and by Cannon and MacGillivray [14]. As far as possible the complete system should be calibrated in the conditions in which it will be used. Particular problems may arise with strain gauge load cells where dynamic calibration is not normally possible, where a.c. carrier wave systems restrict response, where transducer response varies non-linearly with frequency or where the natural frequencies of vibration of the transducers or machine interfere with the measurement. Large errors can arise due to transducer design; in the disc-type Charpy pendulum load cell shown in Figure 13.3 the gauges are fitted into a recess to give protection and place them close to the striking face. This makes the output very sensitive to small changes in the position of contact with the specimen and in-situ calibration must be performed.

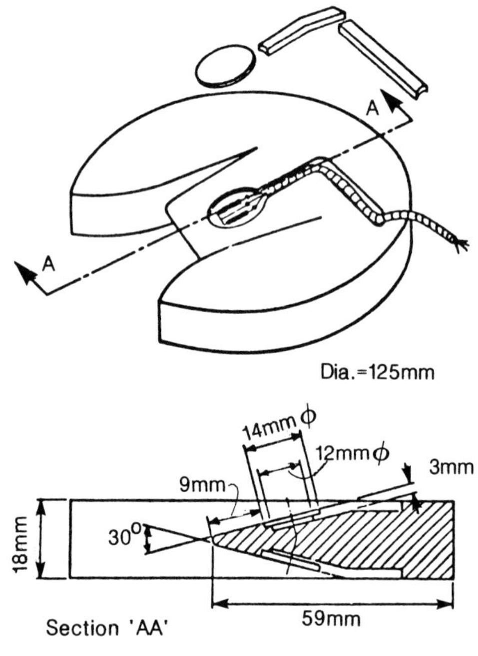

Figure 13.3: The instrumented Charpy pendulum

For pendulum machines a simple overall calibration can be obtained by comparing the energy absorbed by a ductile testpiece, calculated from the area under the load-time record, with the loss in energy measured from the dial. Allowance can be made for any reduction in velocity during fracture when converting from load-time to load-displacement, and agreement to within a few percent is possible. Additional confidence is gained from performing the calibration on a standard Watertown Arsenal aluminium V-notch specimen. An equivalent procedure can be used with dropweight machines by measuring the energy absorbed in deformable stop-blocks.

Accidental or deliberate filtration of signals before recording should be avoided as vital information can be lost. If smoothing is felt necessary it can be performed on the stored data, allowing comparison with the original record.

Piezoelectric loadcells

These are often used in new and converted testing machines. They provide an economical and robust force measurement provided any limitations in calibration and response are considered.

Strain gauges

Both foil and semiconductor types are widely used for force measurement, normally as a full 4-arm bridge loadcell in a protected location close to the striker note. Foil gauges have the advantages of a large choice of geometry, low cost, easy application, stability and reliability. An amplifier is normally required. Semiconductor gauges are usually small with low mass and can be used unmounted so there is no backing to affect strain transmission. Because of their large gauge factor no amplifier is needed. Disadvantages comprise occasional sudden failures of lead-gauge connections and zero drift with varying temperature, though this is not a problem in a laboratory. However, where single semiconductor gauges are used as, for example, by Yokoyama and Kishida [15] on a Hopkinson pressure bar, provision must be made for regular calibration.

The idea of using the signal from a calibrated strain gauge on the specimen to measure the applied load as a means of overcoming the dynamic and inertia problems associated with striker loadcells has been widely investigated. The review by Ireland [16] considered all the possible positions shown in Figure 13.4. Some are unsatisfactory because of dynamic bending effects in the specimens. The crack tip gauge position B has advantages where fracture is mainly elastic since the local elastic strain increases almost linearly with increasing K. One disadvantage is the need to strain gauge and calibrate each specimen. An uncalibrated crack tip gauge can also give a good indication of crack initiation for both brittle and ductile fracture as reported by Glover et al [17] and Bassim et al [18].

The crack-mouth gauge position C is promising as a method of measuring CTOD but requires further development. Strain gauges form the measuring element of the current standard transducer for dynamic CTOD, the clip gauge of the type shown in Figure 13.5. This robust and sensitive full-bridge device with gauges on titanium cantilever arms supported by knife edges on the specimen can be used at displacement rates up to about $0.5 ms^{-1}$. Above this natural frequency effects limit the response. The new British Standard [5] provides guidance on selection and use. Clip gauge designs optimised for higher frequency response are possible, perhaps using the omega loop principle, but the trend is towards non-contact techniques.

The final use for strain gauges in dynamic fracture is in recording crack length and velocity by changes in strain as the

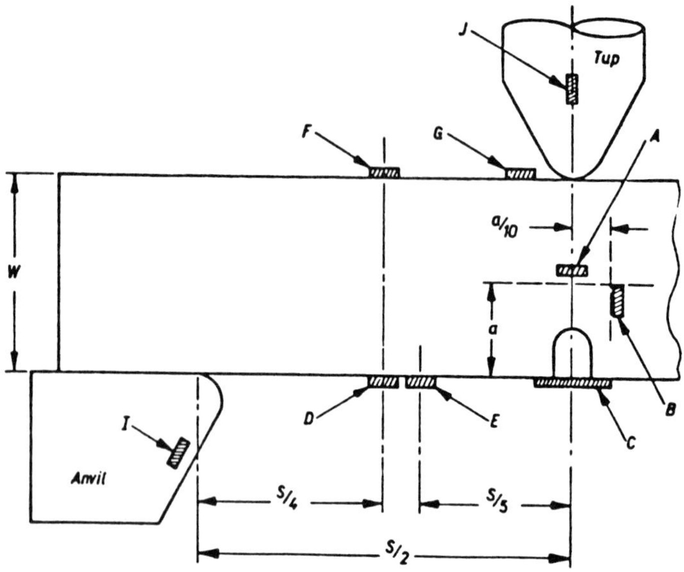

Figure 13.4: *Strain gauge positions*

crack tip passes a row of gauges. This gives good results for thin specimens or cleavage fracture but response may be reduced by ductile shear lips and crack tunneling.

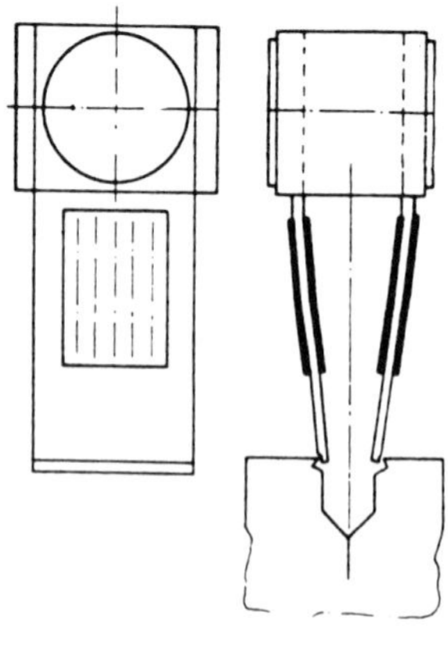

Figure 13.5: *Clip gauges for measuring CTOD*

Encoders

Rotary encoders of various types are often fitted to pendulum impact machines to measure displacement and velocity of the striker. If their frequency response is sufficiently high no problems are encountered unless attempting to measure small changes in velocity. In this case the errors induced by vibration modes in the pendulum described by Kobayashi et al [19] need to be corrected.

Noncontacting displacement measurement

Several precision non-contacting displacement transducers are available with the capacitance type in general use but more suitable for lower loading rates up to about $1ms^{-1}$. Designs are available to fit directly into the machined notch of bend and tension specimens with calibration procedures to correct for angular rotation [20]. Eddy-current devices with high frequency response are now being developed in small sizes suitable for Charpy specimens. These techniques are very promising but they need specialised electronics and put relatively large and fragile transducers on the testpiece.

A novel approach using laser reflection from the V-notch of the specimen as a measure of CTOD has been developed in Finland by Rintamaa and Zimmerman [21] in conjunction with a new design of pendulum impact machine [22]. It gives very clear indication of initiation in the case of limited plasticity, less so when ductile fracture occurs. This is potentially a very useful technique but requires specialised optical equipment and adaption of testing machines.

Laser-Dopler velocimetry has been successfully applied to the measurement of striker velocities up to $2ms^{-1}$ during drop-weight impact testing by Hodgkinson et al [23]. Using light reflected from the side of the falling striker, up to 50 measurements can be made during a fracture event lasting 1ms, allowing energy and toughness to be calculated. With suitable transient recorders, velocities up to $50ms^{-1}$ are possible.

An interferometric method of strain or displacement measurement has been developed by Douglas and Sharpe [24] and applied to CTOD during dynamic fracture. The experimental procedure is based on the motion of fringe patterns reflected from two sets of indents 30 microns square placed 100 and 200 microns behind the tip of the fatigue crack. The technique offers great precision and flexibility in use although in this application surface measurements may not accurately follow the fracture process within the material.

Shadow optic caustics

This technique, developed for dynamic fracture by Kalthoff [25], reflects a beam of light from the concave depression formed at the crack tip as the specimen is loaded. This produces a shadow image called a caustic which can be photographed by a high-speed camera at up to 40000 frames sec^{-1}. The size of the caustic is directly related to the crack tip stress intensity factor, allowing K_{1D} to be determined. The technique has been used to develop impact response curves for linear elastic behaviour, characteristic of any material/geometry combination. Then only a measurement of time to fracture is required, using a crack tip strain gauge or magnetic pickup, to determine K_{1D}. It may be possible to extend this useful independent method into the elastic-plastic regime [26]. The main disadvantage is the need for specialised optical and high speed camera equipment.

Acoustic emission

Investigations by Steffans [27] and at Imperial College [28] on structural steels have shown this method to be a good indicator of cleavage fracture. Ochiai [29] has reported detecting ductile initiation in high-strength steel and aluminium alloys, which if extended to other materials would make this non-contact method very useful.

Direct measurement of crack length

The length of a rapidly growing crack can most easily be measured by a conducting strip or grid on the specimen surface, over an insulating layer if required. A voltage is applied across the crack and as its length increases a continuous or stepwise change in resistance is obtained. The conducting layer can be painted or printed onto the surface, or a commercial crack propagation gauge, comprising a brittle foil and backing, bonded to the specimen. The system works well for limited plasticity or running cracks, but the surface behaviour may not accurately follow ductile initiation and growth where crack front curvature may be significant.

Both a.c. and d.c. potential drop techniques which are used routinely for fatigue crack growth measurement can be applied to dynamic fracture. The a.c. method has limited frequency response and at high frequency lead movement is a source of instability, but d.c. has recently been used for ductile crack growth measurement in steel with some success, as described below.

All the above techniques, either alone or in combination, can give valuable insights into the dynamic behaviour of materials. Two examples of their use are now described in detail.

THE DCPD SINGLE SPECIMEN TEST

Direct current potential drop (DCPD) is in general use for measuring crack length in fatigue and creep crack growth testing of electrically conducting materials. A heavy current is passed through the specimen and the small voltage potential drop across the crack is measured by output leads spot-welded on either side as shown in Figure 13.6. The method has been used for detecting initiation in K_{1D} testing by Varga et al [30]. Curry and Milne [31] reported that it was difficult to reliably detect ductile initiation using this technique. However, recent developments in the use of DCPD at Harwell by Gibson and Druce [32] indicated the possibility of its use in single-specimen dynamic tests if a suitable calibration procedure could be developed. This pro-

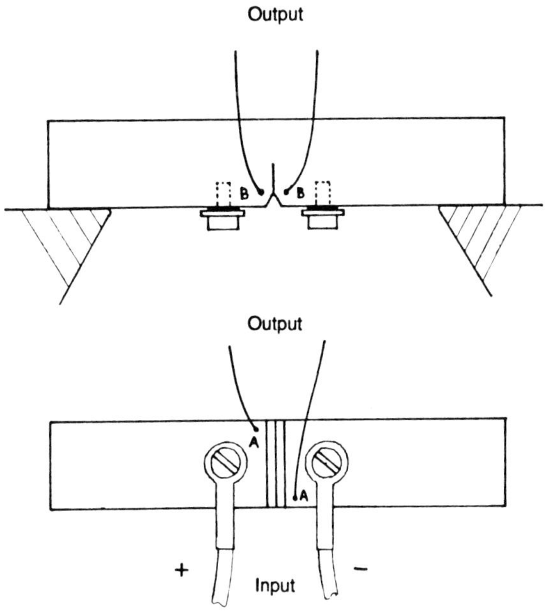

Figure 13.6: DCPD Charpy specimen

gramme, preliminary results of which were reported by MacGillivray and co-workers [33], is eventually intended to perfect a single-specimen test able to measure the variations of performance in small samples cut from welded joints.

Material and test specimens

The specimens used in this programme were machined from Yugoslav manufactured high-strength low-alloy steel equivalent to the US standard HY 100. Standard Charpy dimensions of 10 x 10 x 55mm were used but with 1mm depth V-notch. The specimens were fatigue precracked to BS 5762 requirements to a crack length in the range 0.45 - 0.55a/W. Some specimens were sidegrooved after precracking using a standard Charpy cutter to a depth of 2mm on each side. The potential drop input leads were connected by screws 6mm either side of the notch. Two positions for the output leads were used as shown in Figure 13.6. Input and output lead positions were selected to obtain maximum sensitivity at the expense of some loss of linearity.

Equipment

The DCPD power supply used was Type PDM-1 manufactured by W.H. Mayes. The output was connected to two Fylde 359TA strain gauge amplifiers adapted at D.C. voltage amplifiers with a frequency response of -1dB at 100kHz with gains in the range 10^3 -10^4.

This response and the trigger synchronisation was checked using a function generator.

The fatigue precracking, static and medium strain rate tests were performed in a servohydraulic testing machine. For the high strain rates an instrumented Charpy machine was used with disc striker fitted with a semiconductor strain gauge loadcell, previously described in [7]. This machine has a striker radius of 8mm and a maximum energy of 288 joules at $5.3ms^{-1}$.

For the static tests the PD and load outputs were plotted against crosshead travel on two 2 pen X-Y recorders. In the dynamic tests two Datalab DL905 5MHz transient recorders with 1K memory and 8 bit resolution triggered simultaneously by an infra-red position detector were used to record the load and one PD output against time. The records could be viewed on an oscilloscope and plotted by a X-Y plotter. A block diagram is shown in Figure 13.7.

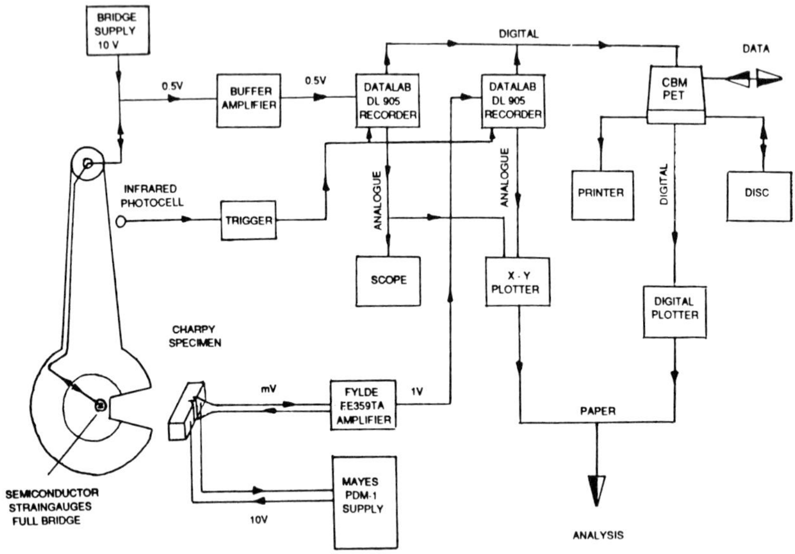

Figure 13.7: Diagram of instrumented impact system

Test programme

Preliminary tests established that the PD output was linear with change of input current up to at least 30A. All tests were performed using a current of 20A which produced an output of about 0.5mV and resulted in an increase of specimen temperature of 10°C. Tests were also made to establish the effect of roller insulation. This was found to make no measurable difference to the PD record. The rollers used for the static tests were normally insulated, but the specimen was in direct contact with the abutments for the impact tests. Specimens were loaded in three-point bending at a rate of 10mm min^{-1} to different amounts of displacement, recording load and PD against ram travel. They were heat-tinted at 250°C for 30 min, cooled in liquid nitrogen and broken open. The fatigue and ductile crack growth was measured at three positions. As a comparison one specimen was tested in fatigue at 10Hz, monitoring the crack length visually and by beach marking at intervals at 20Hz, while continuously recording the PD.

Specimens were tested in the same machine at the maximum displacement rate of $0.184ms^{-1}$. Trials showed that this rate was reached after 3mm travel and remained constant through specimen fracture. In the tests 10mm pre-travel and 10mm displacement was used which ensured complete fracture. The specimens were retained against the abutments of the bending rig by the application of a small amount of cyanoacrylate adhesive. Load and one PD output was recorded against time.

In the Charpy impact tests specimens were tested at $5.3ms^{-1}$. They were again retained by adhesive, and load and one PD were recorded.

RESULTS FROM PD EXPERIMENTS

A typical plot of load and PD against displacement for a static test is shown in Figure 13.8. There is an initial rapid increase in PD corresponding to the separation of the fatigue crack faces. This is followed by a plateau where initiation occurs, then a gradually increasing slope as crack growth continues. In some tests there was a small decrease in PD around initiation. By plotting PD against incremental J measured from the load trace as in Figure 13.9 a sensitive measure of initiation can be obtained, indicating a J_{1C} in the range .075-.090 MNm^{-1} for the side-grooved specimens. A JR-curve was constructed by plotting total J up to unloading against measured final crack length for each of the specimens tested.

In Figure 13.10 the change in PD between the initiation plateau and the final value is plotted against the measured ductile crack growth for all the static test and those dynamic tests in which the specimen remained unbroken. A linear relationship is obtained up to 4mm crack length or 0.9a/w. The calibration obtained from the fatigue tests is also plotted and shows good agreement up to at least 3mm crack growth.

Figure 13.11 shows the type of load and PD trace recorded at $0.184ms^{-1}$. The initial rise in PD is followed by a rapid single oscillation whose cause is not fully understood but which may be a dynamic crack closure effect. After this oscillation the PD continues to rise but is offset from the original zero. There is a small but distinct plateau at initiation followed by a linear increase in PD corresponding to crack growth. By dividing the diagram into sections as shown, the J value calcul-

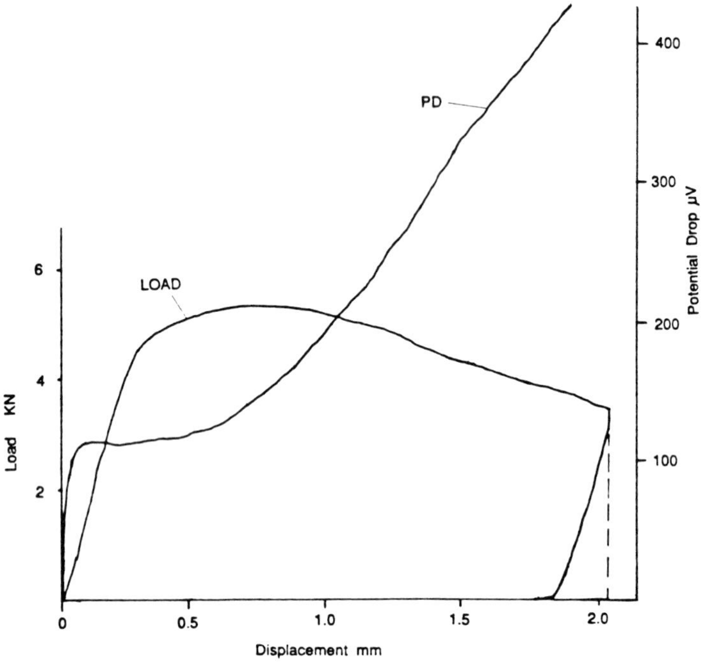

Figure 13.8: *Record of load and PD against displacement*

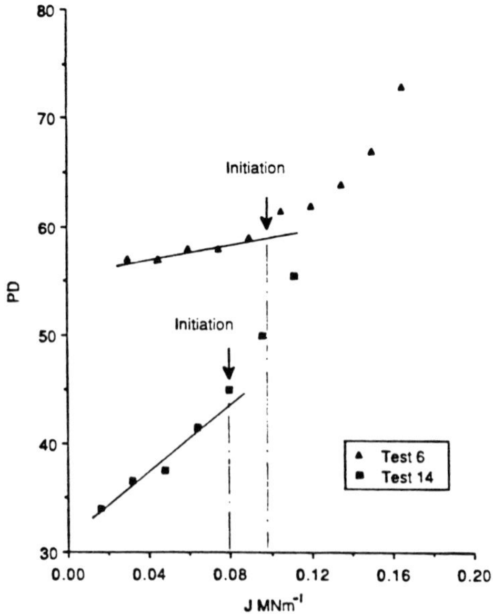

Figure 13.9: *PD versus J integral*

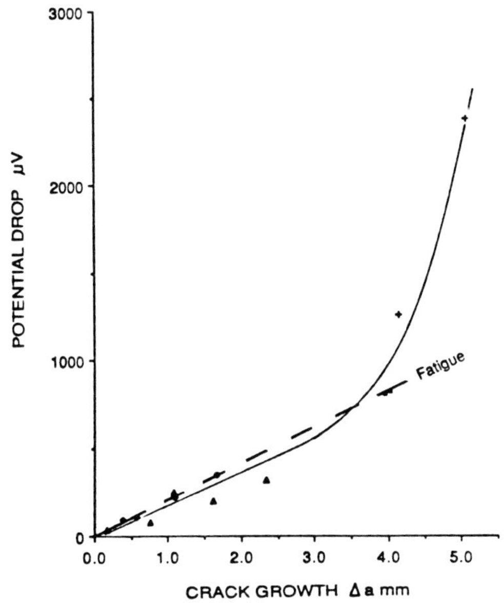

Figure 13.10: *Graph of (final PD - initiation PD) against crack growth*

ated from the load-displacement trace can be associated with a corresponding crack length obtained by using the PD calibration from Figure 13.10. At 5.3ms^{-1} the PD oscillation is more pronounced which makes the initiation point difficult to determine, but the form of the record is similar.

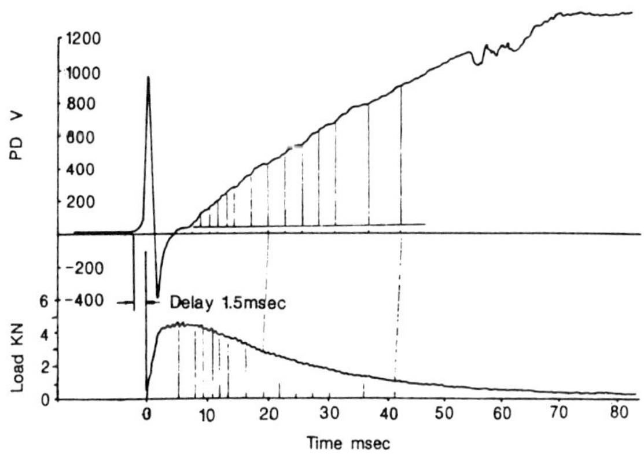

Figure 13.11: *Load and PD record at 0.18ms^{-1}*

JR-curves calculated from the PD data at the three loading rates are shown in Figure 13.12, with additional data points observed from subsequent multi-specimen tests with shouldered specimens of the type developed by Chipperfield [34]. These indicate a marked increase in J_{1D} for the tests at $0.184ms^{-1}$, with no further increase and possibly a decrease at the higher rate of $5.3ms^{-1}$. As expected, the effect of sidegrooves was to reduce the measured J values because of the increase in constraint at the crack tip; sidegrooved specimens also give clearer indication of initiation. Output lead position A-A in Figure 13.6 is more sensitive than B-B. Dynamic initiation toughness J_{1D}, taken directly from the PD or from the PD-J plot is probably not sufficiently reliable at this stage of development. Attempts to obtain unambiguous detection of initiation using crack-tip strain gauges have not yet been successful. The increase in J_{1D} is unexpected, since Bassim [18] found a 50% decrease in tests on a similar HY80 steel using wedge loaded compact specimens of 16mm plate. However, Curry [31] reported an increase in J_{1D} and steeper dynamic R-curves for a BS 1501:271 pressure vessel steel. Similar behaviour has been observed by Bruninghaus [35] in measurements on a ferritic structural steel Fe510.

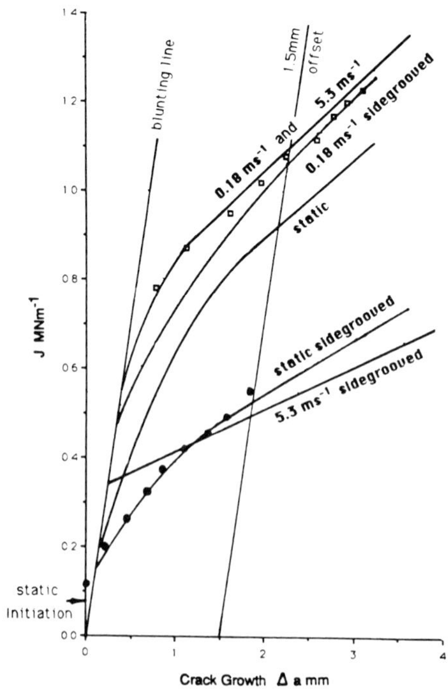

Figure 13.12: JR-curves for static and dynamic tests

DYNAMIC INITIATION TOUGHNESS USING CRACK TIP STRAIN GAUGES

To investigate the possibility of extending the provisions of the new British Standard BS 6729:1987 [5] to higher loading rates, the UK Working Party on Dynamic Fracture recently organised a collaborative test programme among five laboratories. This used the method of recording simultaneously the striker load and the response of calibrated strain gauges, directly bonded close to the crack tip or across the crack tip or across the crack mouth, to indicate initiation. The strain gauge method was chosen on the basis of low cost, known technology and wide application. The results, summarized in Figure 13.13, show reasonable agreement for brittle and ductile behaviour on 10mm x 10mm and 20mm x 40mm SENB specimens.

LAB	CHARPY		20mmx40mm BEND		Strain gauge Locations
	BS 11	50D	BS 11	50D	
A	tests at 0.5 and $5ms^{-1}$		30-50	80-100	crack tip frontface
B	-	39-47	-	-	crack tip back face
C	35	-	-	-	crack tip mouth
D	-	-	47-50	-	crack tip mouth
E	43-46	45-66	-	92	crack tip (2 positions)
F	-	-	21-30	55	crack tip mouth back face

K_D in $MPam^{0.5}$ Tested at $5ms^{-1}$

Figure 13.13: Results of collaborative programme

A comprehensive programme is now in progress at Imperial College to investigate and develop this method and compare it with other available techniques. The first phase has concentrated on optimising the strain gauge and adhesive type, and finding the best gauge position.

A dynamic finite element programme previously described [10] was used to map the strain field around the crack tip. For the SENB geometry this confirmed that a position close to and in line with the crack tip experiences near-linear strain increase up to initiation. Since the first tests were on Charpy specimens, gauges with grid size 1.5mm x 1.5mm were selected to balance the requirements of resolution with availability and ease of handling. The following gauge types were chosen as representing the widest range of foil and backing properties:

standard	EA	constantan + polyimide	-195 + 205°C
large tab	CEA	constantan + polyimide	-195 + 205°C
transducer	MA	constantan + epoxy	-195 + 175°C
wide range	WK	K-alloy + phenolic	-269 + 400°C
tough	WA	constantan + phenolic	-195 + 260°C

The adhesives selected were a standard cyanoacrylate and a thinned epoxy cured at 160°C. Each specimen had two gauges of different types bonded on opposite sides of the crack. The gauges were connected in 3-wire quarter-bridge configuration to Fylde TA 359 amplifiers and Datalab DL 905 5MHz transient recorders. Frequency response was -3dB at 300 KHz with gauge voltage 2v and gain around 100. The strain gauged specimens were calibrated in situ up to the maximum pre-cracking load of 3kN using a proving ring bearing against the striker. The specimens were retained against the machine abutments by light springs to ensure good alignment.

A typical test record for the bittle fracture, in this case BS11 carbon-manganese rail steel tested at 20°C and 5.3ms^{-1}, is shown in Figure 13.14. The load-time trace for the strain gauge shows the expected linear increase to a clear point of fracture. Using a 5% offset procedure, the load at fracture P_c can be found unambiguously. The striker load record shows no indication of the fracture event, while the maximum gives a fracture load around double the strain gauge value. Time to maximum load is also different. Results from a number of similar tests are plotted in Figure 13.15 as K_D against time to fracture t_f. These results at 20°C show no obvious variation due to gauge type but there is a marked reduction in scatter for the epoxy adhesive, probably due to the more uniform glue line thickness achievable.

This procedure has been applied to a practical case with interesting results. The material was En 28 type Ni-Cr-Mn steel, some batches of which were giving service failures under high-rate dynamic loading. Standard blunt-notch Charpy tests showed no obvious differences. Fatigue pre-cracked and strain gauged specimens approximating to Charpy geometry were tested and showed no variation between batches at loading rates up to 1ms^{-1}. However, at 5.3md^{-1} the K_D of the defective material was reduced to 26% of normal.

Only limited success has been achieved using strain gauges for ductile fracture. The load-time records are of the form shown in Figure 13.16, and while the point marked MAX on the strain gauge trace probably corresponds to crack initiation, the connection is less clear. Current work is aimed at investigating the brittle-ductile transition region by testing at low temperatures, and eliminating the effect of the shear lips by side-grooving the specimens. Comparative tests using the shadow optic caustic technique are also in progress.

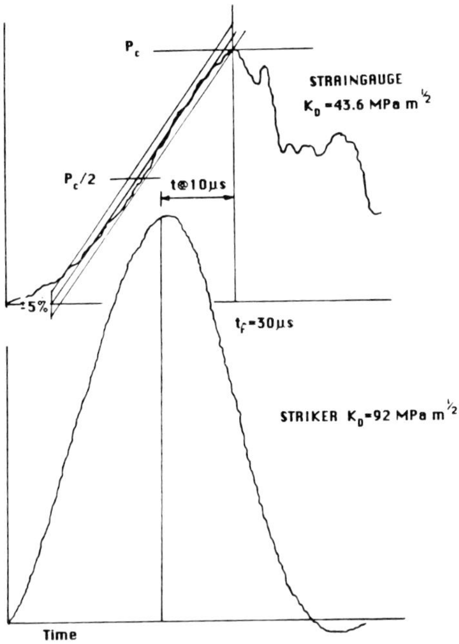

Figure 13.14: A typical brittle fracture record

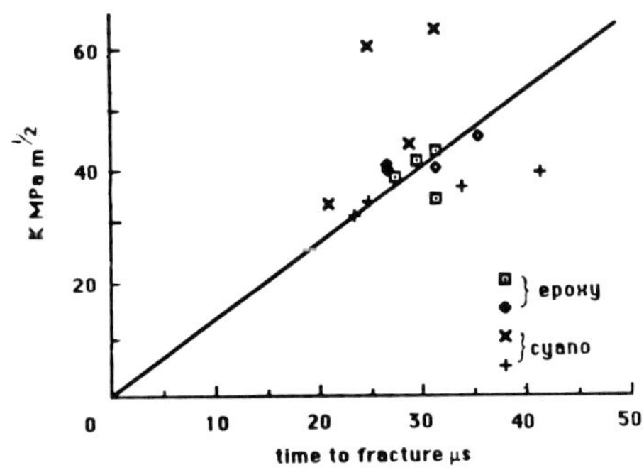

Figure 13.15: K versus time to fracture

CONCLUSIONS

The very wide range of measuring techniques which have been applied to the problems of dynamic fracture have been examined.

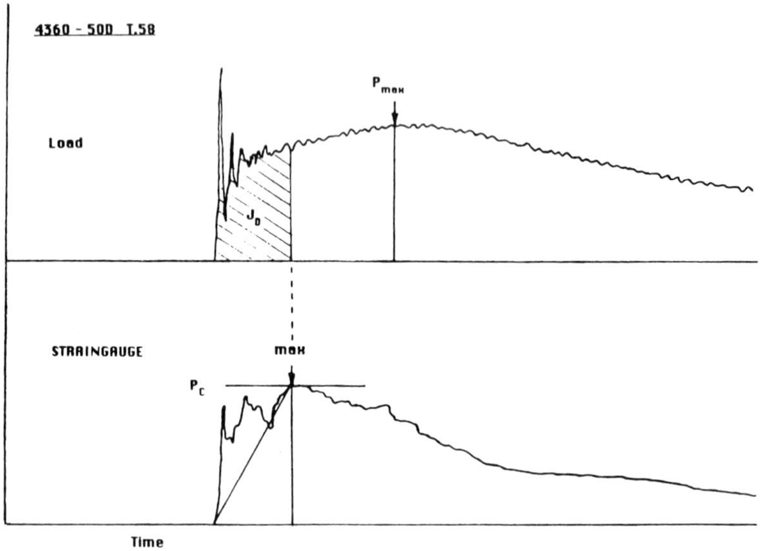

Figure 13.16: A typical ductile fracture record

Each has advantages and drawbacks, and many are complimentary. As yet there is no single method applicable to all situations, nor is there any British or International standard for very high rate testing of this type.

Since strain gauges form the basis of so many of the methods, there appears to be a need for a high frequency dynamic gauge calibration procedure.

ACKNOWLEDGEMENTS

The authors would like to thank members of The UK Working Party on Dynamic Fracture for their advice and for permission to use results, and the Department of Trade and Industry for financial support.

REFERENCES

[1] ASTM E399-83"Standard Test Method for Plane-Strain Fracture Toughness of Metallic Materials", American Society for Testing and Materials, Philadelphia (1986)
[2] ASTM E 813-81 "Standard Test Method for J_{1C} a Measure of Fracture Toughness", ASTM, Philadelphia (1986)
[3] BS 5447:1977, "Methods for Plane-Strain Fracture Toughness (K_{1C}) Testing", British Standards Institution, London
[4] BS 5762:1979, "Methods for Crack Opening Displacement (COD) Testing", British Standards Institution, London

[5] BS 6729:1986, "Method for the Determination of the Dynamic Toughness of Metallic Materials", British Standards Inst., London

[6] Tanaka, M. and Umekawa, S., "On the Breaking Behaviour in Charpy Impact Bending Terms", Proceedings, 1st Japan Congress for Testing Materials, 95 (1958)

[7] Radon, J.C. and Turner, C.E., "Fracture Toughness Measurements by Instrumented Impact Test", Journal of Engineering Fracture Mechanics, 1, (3), pp. 411-428 (1969)

[8] Server, W.L. and Tetelman, A.S., "The Use of Precracked Charpy Specimens to Determine Dynamic Fracture Toughness", Eng Fract Mech, 4, 367 (1972)

[9] Kobayashi, T., Yamamoto, I. and Niinomi, M., "Evaluation of Dynamic Fracture Toughness Parameters by Instrumented Charpy Impact Test", Eng Fract Mech, 24, (5), pp. 773-782 (1986)

[10] Curr, R.M., MacGillivray, H.J. and Turner, C.E., "A Comparison of Some Elasto-plastic Finite Element, Experimental Data and Some Simple Models for Notched Bars and Plates", IMechE Seminar on Dynamic Fracture Mechanics, London (1986)

[11] Kobayashi, A.S., "Dynamic Fracture Analysis by Dynamic Finite Element Method", Non-linear and Dynamic Fracture Mechanics, ed. Perrone and Atluri, ASME AMD 35, pp. 19-36 (1979)

[12] ASTM 24.03.03, "Proposed Method of Test for Instrumented Impact Testing of Precracked Charpy Specimens of Metallic Materials", American Society for Testing and Materials, Philadelphia (1981)

[13] McConnell, P. and Server, W.L.m "EPRI Instrumented Impact Test Procedures", CSNI Specialists Meeting on Instrumented Precracked Charpy Testing, Palo Alto, Calif. (1980)

[14] Cannon, D.F. and MacGillivray, H.J., "A New British Standard for the Determination of the Dynamic Fracture Toughness of Metallic Materials", Int. Conf. on Impact Loading and Dynamic Behaviour of Materials, Bremen (1987)

[15] Yokoyama, T. and Kishida, K., "Measurement of Dynamic Fracture Initiation Toughness by a Novel Impact Three-Point Bend Test Technique using Hopkinson Pressure Bars", Int. Conf. on Impact Loading and Dynamic Behaviour of Materials, Bremen (1987)

[16] Ireland, D.R., "Critical Review of Instrumented Impact Testing", Welding Institute/ASM Int Conf on Dynamic Fracture Toughness, 1, pp. 47-62 (1977)

[17] Glover, A.P., Johnson, F.A., Radon, J.C. and Turner, C.E., "Dynamic Fracture Toughness Measurement by Instrumented Impact Bend Testing and Compact K Testing", Welding Institute/ASM Int Conf on Dynamic Fracture Toughness, 1, pp. 63-75 (1977)

[18] Bassim, M.N., Bayoumi, M.R., Hsu, T.R. and Matthews, J.R., "Investigation of Dynamic J_{1D} for Alloy Steel Weldments using the Split Hopkinson Bar", Journal of Testing and Evaluation, 14, (5), pp. 229-235 (1986)

[19] Kobayashi, T., Yamamoto, L. and Niinomi, M., "On the Accuracy of Measurement of Dynamic Elastic-Plastic Fracture Toughness Parameters by the Instrumented Charpy Test", Engineering Fracture Mechanics, 26, (1), pp. 83-94 (1987)

[20]Foster, R.L., "A New Method for COD Measurement in Fracture Testing", Seminar on Advanced Instrumentation for Experiments in Engineering Mechanics, Knoxville (1984)

[21]Rintamaa, R. and Zimmermann, C., "Advanced Instrumented Impact Testing Facility for Characterisation of Dynamic Fracture Behaviour", 11th MPA Seminar (1985)

[22]Rintamaa, R. et al., "Instrumented Impact Testing Machine with Reduced Specimen Oscillation Effects", Research Report 290, Technical Research Centre of Finland (1984)

[23]Hodgkinson, J.M., Vlachos, N.S., Whitelaw, J.H. and Williams, J.G., "Drop-weight Impact Tests with the use of Laser-Doppler Velocimetry", Proceedings of the Royal Society, London, A 379, pp. 133-144 (1982)

[24]Douglas, A.S. and Sharpe, W.N., "Fracture Initiation under Impact Loading", Int. Conf. on Impact Loading and Dynamic Behaviour of Materials, Bremen (1987)

[25]Kalthoff, J.F., "On the Measurement of Dynamic Fracture Toughnesses - a Review of Recent Work", International Journal of Fracture, 27, pp. 277-298 (1985)

[26]Hermann, R., The Open University, (private communication)

[27]Steffans, H.D., Crostack, H.A. and Engelhardt, A.H., "The Application of Acoustic Emission Analysis to the Determination of Material Behaviour under Dynamic Loading", Acoustic Emission (Proc. Conf.), Bad Nauheim, W. Germany, pp. 205-216 (1980)

[28]Oldroyd, P., MacGillivray, H.J. and Turner, C.E., N11 Report on Instrumented Charpy type Impact Tests on A533B Steel (1983)

[29]Ochiai, S.I., Cheresh, M.C. and Green, J.E., "Acoustic Emission Detection of Crack Initiation during Dynamic Fracture Testing of High-Strength Materials", J. Acoust. Emiss., 2, (4), pp. 289-291 (1983)

[30]Varga, T., Njo, D.H. and Prantl, G., "ASK Procedure for Instrumented Precracked Charpy-Type Tests", CSNI Specialists Meeting on Instrumented Precracked Charpy Testing, Palo Alto, Calif. (1980)

[31]Curry, D.A. and Milne, I., "The Detection and Measurement of Crack Growth during Ductile Fracture", Symp on Measurement of Crack Length and Shape, Birmingham, pp. 401-434 (1980)

[32]Gibson, G.P. and Druce, S.G., "Development of the Direct Current Potential Drop Technique for Determining J-Crack Growth Resistance Curves", AERE Report R11443, Harwell (1984)

[33]MacGillivray, H.J., Grabulov, V. and Akum, E.R., "Development of the DC Potential Drop Method for Dynamic J-R Curve Testing of Charpy Specimens", Inst Symp on Welding and Testing of Metal Stuctures in Civil Engineering, Belgrade (1987) (in Serbian)

[34]Chipperfield, C.G., "A Method for Determining Dynamic J_Q and d_i Values and its Application to Ductile Steels", Welding Institute/ASM Int Conf on Dynamic Fracture Toughness, 1, pp. 167-179 (1977)

[35]Bruninghaus, K., Hesse, W., Twickler, M. and Dahl, W. "Effect of Prestraining on Fracture Behaviour of Fe 510 Steel under Dynamic Loading", Int. Conf on Impact Loading and Dynamic Behaviour of Materials, Bremen (1987)

14 Interferometric systems for measuring crack-opening displacements of small cracks

W.N. Sharpe Jnr

SUMMARY

Very small fatigue cracks often grow faster than would be predicted from the behaviour of longer cracks. Some insight into this phenomenon can be gained from studies of the crack-opening displacement as a function of applied load, but this is difficult on such a small scale. In-plane interferometry in which tiny indentations are placed across the crack and illuminated with a laser to produce fringe patterns can make these measurements with suitable resolution. Three versions of computer-based interferometric measurement systems that are currently used for this difficult measurement problem are described in this chapter.

INTRODUCTION

There is ample evidence that very small fatigue cracks (less than 1mm long, say) can grow faster than would be predicted from knowledge of longer cracks. This phenomenon, referred to as the "small crack effect", has come under scrutiny in the past decade because of the increasing resolution of nondestructive evaluation (NDE) techniques. There are certain critical components or systems where the time and expense of high-resolution NDE examination for small cracks is justified; an example is a rotor disk of an aircraft jet engine. Small crack effects continue to be an active area of research, and the proceedings of a conference held at the beginning of 1986 [1] presents some of the issues involved.

One possible explanation of the small crack effect is that the closure associated with small cracks may be quite different than for longer cracks. It is well-known from fatigue studies that a crack does not necessarily open all the way to its tip immediately upon the application of a non-zero load, but may very well open at a higher load, P_{op}, which is a significant fraction of the maximum applied load. P_{max}, (this phenomenon is called closure). This being the case, the actual load range applied to the specimen or component must be recalculated and this calculation may differ between small and long cracks. If closure does explain the differences (and it is not firmly established that it does), it is then necessary to determine the opening load ratio, P_{op}/P_{max}, over a range of crack lengths, loading ratios, materials, etc.

There are no standard ways of measuring the opening load ratio for long cracks, but a very common approach is to measure the crack-opening displacement as a function of the applied load. This curve will show a linear region for the larger loads when the crack is fully open; before then, the curve is nonlinear because the crack is "peeling" open. If this approach is followed for small cracks, then one must obviously be able to measure the crack-opening displacement on a very fine scale with high resolution.

To permit study small cracks must be generated, located and then measured in laboratory specimens. None of these are trivial problems because one really likes to start the study with cracks that are on the order of tens of micrometers long. After the cracks have been produced, then the crack-opening displacement can be measured with the laser-based Interferometric Strain/Displacement Gauge (ISDG). Tiny indentations are placed across the small crack with a Vickers micro-hardness tester. When illuminated with a laser, they generate interference fringes in space whose motion is proportional to the change in distance between them. A minicomputer-controlled measurement system for monitoring the fringe motion enables a resolution of 0.01 μm with a range of tens of micrometers. Indentations as small as 10μm can be placed across fatigue cracks which are only 50 μm long on the surface, and quite respectable load-displacement records can be obtained. From these, P_{op}/P_{max} can be obtained for a variety of loading conditions and crack lengths.

This chapter is an overview of three minicomputer-controlled ISDG systems that are currently in use for small crack studies. First, the acetate replica technique for finding small cracks is described. Then, three approaches are described: an incremental-sweep scanning mirror, a single-sweep scanning mirror, and a linear array detector. Typical results and a discussion of the relative advantages and disadvantages of the three approaches conclude the chapter.

ACETATE REPLICAS

Searching for cracks 10 to 20μm long can be very tedious, so the initiation area must be localized. This can be done with notches or other stress concentrations, but that is not entirely satisfactory because the stress field complicates the basic behaviour of the cracks. Specimens with small stress concentrations, of the order of 1.1, are often used, and the experimenter simply has to search over a fairly large area. Another approach that works in aluminium is to lightly etch a very small region of the specimen with Keller's etch; cracks tend to initiate within this small area.

Acetate replicas are straightforward, easy-to-use records of the microstructure of a metal (or other material) surface. The replica material is acetylcellulose from Bioden RFA and is 34μm thick. This is a special, high quality material, but successful replicas have been taken with the acetate material used for overhead transparencies. A drop of acetone is placed on the specimen surface, and then the replica material is pressed over it with thumb pressure for approximately one minute. Surface tension will then hold the material in place where it is allowed to dry for five minutes or more. The acetone dissolves the acetate into the surface features and produces a replica of high fidelity. The replica can then be examined with an optical microscope or, if desired, with an SEM. A thin, vapour-deposited coating can enhance the contrast for optical microscopy (a conducting coating is required for an SEM), but the bare replicas are quite satisfatory. There are several advantages to using replicas:

- a specimen surface can be examined at high magnification without removing it from the test machine
- locations that are difficult to access can be examined
- a permanent record is created
- one can "look back" - often after a small crack has been found, one can look at earlier records and find that it was visible earlier, but was overlooked.

In a recent series of experiments in 2024-T3 aluminium sheets, measurements were made of the crack size with the acetate replica technique, indents were applied for the load-displacement measurements, and then the cracks were grown until they crossed the specimen thickness and grew down the sides of the specimen. Indentations can be placed at various positions along the crack as it grows; it is necessary that a set of indents be more than ~300μm apart to avoid having the laser beam illuminate more than one set.

BASICS OF THE ISDG

Figure 14.1 is a photomicrograph of a pair of indentations impressed with a Vickers microhardness tester across a small crack in the specimen surface. The indents are 60 μm apart and are each approximately 20μm square and 4μm deep.

When an indentation is illuminated with a laser beam perpendicular to the specimen surface, light is diffracted from each of the four triangular-shaped sides. Since the indents are so close together, the diffracted light rays overlap and interfere to form fringe patterns in space. The position of fringes of a given order in the patterns above and below the set of indents (as viewed in Figure 14.1) is proportional to the distance between the indents. This is simply the two-slit interference phenomenon of Young [2] except that it is in reflection, not transmission. The other two fringe patterns on either side of the indentations are not used in these measurements.

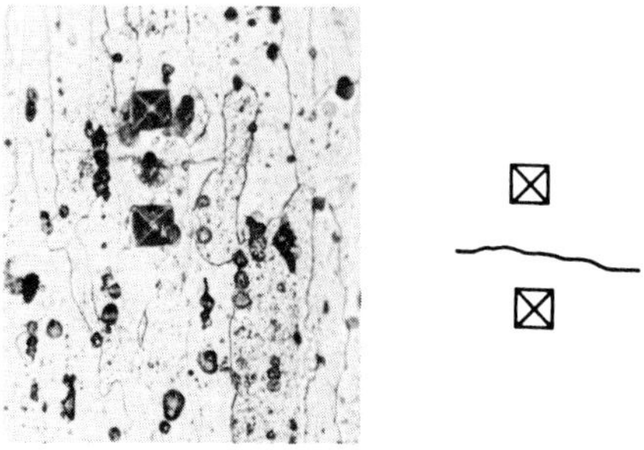

Figure 14.1: *Photomicrograph of a set of indentations across a small surface crack. The centres of the indents are 60μm apart, and the crack is 85μm long. The schematic to the right shows the position of the indents and crack*

It is the relative change in the distance between the indents that is of interest here, and that change causes motion of the fringes. The interference fringes move within the diffraction pattern emanating from the individual indents. More details of the optical principles are given in [3]. The relation between the fringe motion and the relative displacement is given by:

$$\delta d = \frac{\delta M_1 + \delta M_2}{2} \frac{\lambda}{\sin \alpha_0} \tag{14.1}$$

where λ is the wavelength of the laser, α is the angle between the incident laser beam and the fringe patterns, and $\delta M_1, \delta M_2$ are the relative fringe motions. "Relative fringe motion" means the displacement of a fringe divided by the distance between it and the next fringe. It is necessary to average the two relative fringes shifts to eliminate the effects of rigid body motion in the vertical (in Figure 14.1) direction.

The wavelength of a He-Ne laser is 0.628μm, and α_0 is typically 42 degrees. The quantity $\lambda/\sin\alpha_0$ is then 0.94μm; this means that an average relative fringe motion of 1 corresponds to a relative displacement of that amount. There are many applications where a resolution of that order is quite satisfactory, but it is really too coarse for the crack opening displacements associated with small cracks.

The basic resolution of the ISDG is ~1/2 μm - corresponding to a fringe motion from bright to dark. If finer resolution is required, then some means of measuring fractional fringe motion (or interpolating between fringes) must be used. There are many schemes for doing this with a static pattern such as found in photoelasticity or in full-field interferometry. However, the ISDG is not a full-field technique, and one needs to measure the fractional fringe motion on a realtime (or nearly realtime) basis. It may be possible to do this in an analogue fashion, but three computer-controlled systems, based on different optical sensors and strategies, have been developed to-date: an incremental-sweep scanning mirror [4], a single-sweep scanning mirror [5], and a linear-array detector [6]. In all three systems, the electrical analogue of the position of the fringes is acquired at a rate of much faster than the motion of the fringes, i.e., a "snapshot" of their position is taken.

The fringe pattern is shown schematically in Figure 14.2; the optical system would be adjusted so that the electrical signal proportional to the intensity would correspond to a sweep over three minimums. The approach is to follow, on a realtime basis, the movement of a maximum or minimum as it moves due to the relative displacements of the indents. This motion, when divided by the spacing between fringes, produces the fractional fringe motion to be used in Equation 14.1. The incremental-sweep approach samples the fringe pattern only between B and D in Figure 14.2 and tracks the motion of the minimum at C. The single-sweep and linear array detector approaches sample the fringe pattern from A to E, but still follow the motion of the minimum at C. Of course, if C moves more than one fringe spacing, the tracking strategy must be able to shift to an adjacent minimum that is near the centre of the sweep.

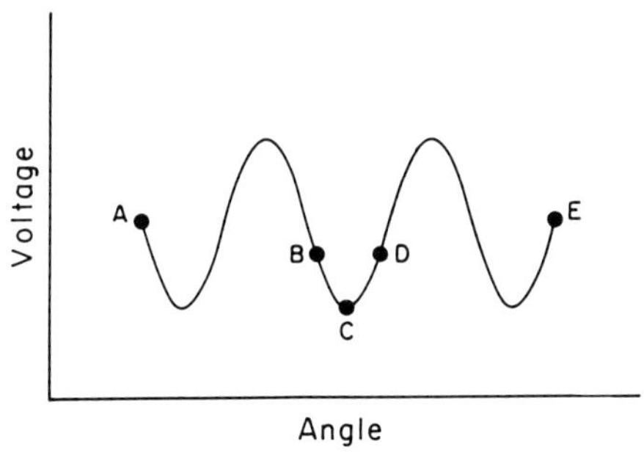

Figure 14.2: Schematic of a 'sweep' of three dark fringes

INCREMENTAL-SWEEP SCANNING MIRROR

Figure 14.3 is a photograph of the scanning mirror system in use at Johns Hopkins. The fringe patterns emanating from the specimens are directed onto photomultiplier tubes by the scanning mirrors which are controlled by the minicomputer. The two mirrors, the two photomultiplier tubes and the laser are mounted on a vertical aluminium plate that is attached to the test machine. A steering mirror is used to turn the laser beam 90 degrees and aim it precisely onto the indentations. The mirrors are mounted on stages that translate them perpendicular to the fringe pattern and rotate them about their scanning axis. The adjustments are needed to locate the scanning mirror in the fringe pattern emanating from the specimen.

Figure 14.3: Photograph of the scanning mirrors, photomultiplier tubes, laser, and steering mirror mounted on a plate attached to the testing machine. The specimen in this case is a compact tension one with a back-face foil strain gauge attached

The scanning mirrors are produced by General Scanning, Inc. of Watertown, MA and are model G120DT. These are torsional shaft scanners, i.e. one end of the small shaft is fixed and a small mirror (7mm by 16mm by 1mm thick) is fastened to the other end. A magnetic core is attached to the rod, and the electric field in a surrounding coil controls the position of the mirror. A capacitance-based angular transducer is built into the system and provides the feedback signal for control. The model CX-660 control module is used (one for each mirror) and transforms a ± 10 volt input signal into the corresponding angular motion of the mirror. A typical sweep is on the order of 1 degree, and the frequency response of the scanning mirror system is approximately 600hz.

The photomultiplier tubes are RCA type 4840 and are powered by RCA model PF 1042 power mudules that require 13.5v DC input. The advantages of these tubes are that they have a high maximum current, a large aperture and are relatively inexpensive.

In a normal setup the mirror is placed approximately 30cm from the specimen which causes the 16mm length to intercept three fringes. The control unit has an offset control which is used to centre the pattern and a gain adjustment which controls the magnitude of the angular sweep. A computer program that produces a sawtooth waveform must be used to sweep the mirrors during initial setup while one observes the fringe pattern signal on an oscilloscope that is triggered externally (the trigger signal is generated by the program also).

The computer used in this version at Hopkins is a Digital Equipment MINC laboratory minicomputer which, unfortunately, is no longer produced. It comes with two 7-inch flexible disk drives; one adds the D/A and A/D modules. The configuration used has four channels of D/A and eight channels of A/D plus a 20Mb hard-disk drive. The throughput for the A/D is 25khz. All programs are written in FORTRAN, but the callable subroutines for D/A and A/D were written locally in assembly language.

The 'incremental-sweep' version of the fringe measuring algorithm sends out a signal to move the mirror, takes in a voltage from the PMT, and repeats this process for a specified number of increments. After the data has been acquired, it is processed to locate the position of the minimum in question which is then compared with the minimum location of the previous sweep. The shift, if any, if converted into a relative fringe shift using the original fringe spacing that was determined in the 'setup' phase of the experiment. A signal is then sent to the testing machine to change the load on the specimen and the fringe measuring cycle is repeated.

Because of the time taken for each increment of the sweep, the entire pattern is not sampled; only a region (B to D in Figure 14.2) about the minimum being followed. This approach "locks in" on a particular minimum and follows it as long as it is within the scan of the mirrors. When the minimum of interest moves near the edge of the mirror, the algorithm resets the mirror and searches for a new minimum in a vicinity predicted by the original fringe spacing. The entire fringe pattern is scanned in 256 increments (a holdover from an earlier system that used assembler language on a less powerful computer), but only 40 increments are used to scan around the minimum. The spacing between fringe minima in Figure 14.2 is usually taken to be approximately 100 increments, so one should expect to resolve 1/100th of the fringe spacing which corresponds to a relative displacement resolution of approximately 0.01 micrometers.

A few words about the fringe measurement algorithm may be useful. It is written in FORTRAN as a subroutine that is included in the master program that controls the load and measurement sequence. It has not changed over the past seven years and has been used in a variety of testing programs to measure strain, crack opening displacement, and creep at room and elevated temperatures. First, the two mirrors are incrementally swept over the minimum region and the intensity voltages stored temporarily in an array. These values are then digitally filtered by a sliding 10-point average and placed in another array. This array is passed through with a simple routine that locates the minimum when the values stop decreasing and start increasing. The subroutine identifies the minimum as an element in the array and returns this value to the main program where it is compared with the previous minimum location, converted into relative displacement (with sign) and stored along with load and any other measured values before the next increment of load is applied (or time passes if it is a creep test).

The total time to make one relative displacement measurement, i.e. run through the subroutine once, is 95-100 milliseconds. Roughly half this time is taken in acquiring the fringe intensity data with the other half going to the computations associated with locating the minimums, etc. Low-cycle fatigue tests have been run using 60 points per loading cycle; this translates to a testing frequency of 10 cycles/minute which is acceptable. It is recognized that the subroutine could be speeded up by writing it in assembler language, but it is easily understood by a variety of students and other users in its present form. Even if it were fast enough to follow a typical high-cycle fatigue testing rate of 20hz, it is unlikely that one would want to retain the data on every cycle - not to mention the memory required. For high-cycle fatigue studies, the master program is designed to stop after a certain number of cycles and record a load-displacement or load-strain cycle.

An advantage of this approach is the high degree of stability gained by the precise register of the mirror positions because of the computer control - as opposed to having them swept by an external function generator in the following section. In "fake" creep tests at 650°C (i.e. no load was applied to the heated specimen), the measurement system drifted only ±0.02 micrometers in 1 hour. A disadvantage is that the recorded displacements must be corrected to allow for the change in fringe spacing if they become larger than a few micrometers. The correction formula is [7]:

$$\Delta X = X_0(e^{\delta X/X_0} - 1) \qquad (14.2)$$

where ΔX is the relative displacement of indents that were originally X_0 apart, and δX is the uncorrected relative displacement. When this is used in the main program, excellent agreement between measured displacements on a compact tension specimen and those predicted by linear elastic fracture mechanics was obtained.

SINGLE-SWEEP SCANNING MIRROR

The DEC MINC laboratory minicomputer is no longer manufactured, so when another system was constructed in late 1985, an IBM PC/XT was chosen because of its widespread availability. This minicomputer is not as fast in its A/D and D/A input/output which makes the incremental sweep approach slower. However, the direct memory access (DMA) feature can be used to scan the entire pattern instead of a local minimum and maintain the same sampling rate.

The scanning mirrors and controllers used in this system are exactly the same as in the earlier version; the only difference in the hardware is the minicomputer. The data acquisition board used with the PC/XT is from Data Translation, Inc. in Marlborough, MA. It is model DT 2801-A with an A/D rate of 27.5khz and a D/A rate of 29.5khz. The PCLAB software was purchased from the same company, and all programming was done in compiled BASIC.

The strategy here is to use the triangular waveform produced by one of the CX-660 controllers to drive the other controller and sweep both mirrors simultaneously in a continuous fashion. The controller puts out a trigger signal at the beginning of each positive-going portion of the waveform, and this triggers the minicomputer to make a fringe-motion measurement. Three channels of DMA A/D input were used - one for each fringe pattern and one for the load signal. One channel of regular A/D was used for the trigger signal, and one channel of D/A sent the load signal to the test machine.

Figure 14.4 will help in the explanation of the fringe-motion measurement. The upper part is a plot of one of the PMT signals as the triangular waveform drives the mirror to sweep the fringe pattern across the PMT slit. The lower schematic illustrates the timing involved. At the start of the waveform, the regular A/D channel is sampling the trigger signal repetitively and comparing it with a preset value. When this trigger level is exceeded, the minicomputer sends out a load increment (stored in an array in the "setup" part of the program) to the test machine. This output is immediately followed by the DMA acquisition of the fringe intensities. Both patterns are sampled alternately and continuously to take 512 data points in approximately 30 milliseconds; the odd-numbered data are for one pattern and the even-numbered for the other. This data is temporarily stored in an array while it is processed to locate the minimums.

There are 256 data points for each channel, but the first 50 are not used because the test machine must respond to the small step input and come to equilibrium. Three minimums are located in the same manner as for the incremental-sweep system. The central minimum is followed until it gets too far from the centre of the array; the algorithm then redefines a new central minimum depending on the direction of the motion of the minimums. Relative fringe motion is obtained by dividing the motion of the minimum by the current spacing between minimums. For example, if the original positions of the minimums were array locations 150, 250, and 350, the next scan might show them at locations 154, 254, and 354. This would correspond to a relative fringe motion of 4/100. Of course, in actual practice, the spacing would not be exactly equal nor would the motion of the minimums be exactly equal; this contributes to the "noise" on the output signal. This computation and the conversion to relative displacement and storage on the hard disk takes approximately 30 milliseconds for both channels.

After the computation is completed, a D/A signal corresponding to the relative displacement is sent to an oscilloscope. It is very convenient to monitor the load-displacement behaviour as the test is in progress so that decisions can be made about the quality of the data. There is a "slack" period after the displacement output before the next scan starts. The computation time varies a bit depending upon whether or not a shift is made to a new minimum, and one must allow for this. The frequency of the triangular wave from the controller is set so that the period of one scan is 75 milliseconds; corresponding to a sampling rate of 13 samples per second.

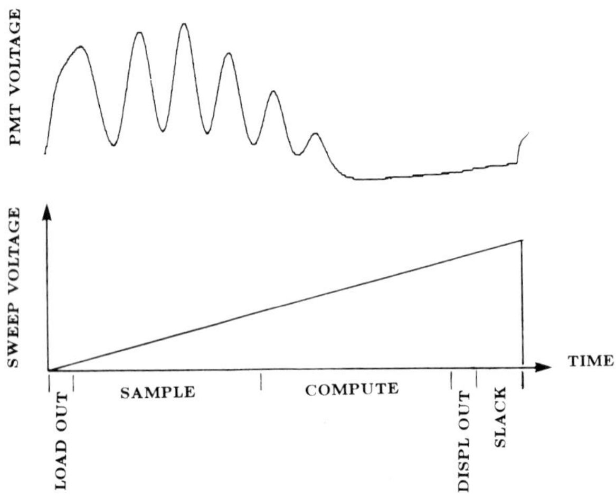

Figure 14.4: *Schematic of the single-sweep scanning mirror approach. The upper part is a typical intensity trace recorded from a photomultiplier tube. The lower part shows the ramp signal sent to the scanning mirror and the timing of the computer operations. The voltage scales are arbitrary, but the period of the scanning ramp is 75 milliseconds*

A distinct advantage of this approach is the continual updating of the fringe spacing in the relative fringe motion computation. The correction for fringe spacing becomes important for displacements larger than roughly 5 µm, so this advantage is not terribly important for small crack studies. An important advantage is that the minicomputer hardware is considerably cheaper. A disadvantage is that the test and measurement is controlled by the triangular wave from the mirror controller which leads to less long-term stability. The function generator must be stable, and the triggering must be very precise. In practice, this stability becomes important only in creep measurements.

LINEAR ARRAY DETECTOR

The third approach is described in a recent article by Hartman and Nicholas [6]. Linear array detectors are used instead of the scanning mirrors and PMT's, and a more powerful minicomputer is used.

The linear array detector, Fairchild model CCD133A, contains 1024 CCDs (charge coupled devices) spaced 13μm apart on a single chip. The aperture for each CCD is 13μm square. This is small enough to be of the order of the speckle size of the fringe pattern, but enough data points are taken and filtered electronically so that this is not a problem. The electronics associated with each array scan each CCD in a sequential fashion and sends its analogue output to the minicomputer through signal conditioning circuits. A trigger signal is produced at the start of each scan to synchronize the data acquisition and the control of the test machine. Each array detector is mounted on a stiff baseplate in front of the test machine and can be adjusted in space to capture the fringe pattern which shines directly onto the detector.

The minicomputer is an IBM System 9000 unit with a 16-bit internal data bus. The A/D converters are 8-bit, and data from both array detectors are taken into the minicomputer simultaneously. The A/D, D/A, and control circuitry were constructed by the developers [6] and are not part of the System 9000.

Each "snapshot" or scan of a fringe pattern consists of 500 data points - not the 1024 possible - simply because of memory limitations of the computer. A complete load-unload cycle has 500 scans. The approach is to run the complete load cycle, acquire the 500,000 fringe data points (2 fringe patterns at 500 points per scan times 500 scans) plus the 500 load values, and post-process the information. The processing algorithm is written in FORTRAN and locates maxima by comparing the largest value within a set of data points with the largest values of two neighbouring sets. The fringe nearest to the centre of the array is followed, and the spacing is updated for each scan. Final output is a plot on a printer at the end of the load-unload cycle.

Data acquisition is at the rate of 200 points per second, i.e. a load-unload cycle takes 2.5 seconds. The post processing takes approximately 7.5 seconds, so a complete cycle from start to finish (viewing the hard-copy output) takes 10 seconds.

A very important advantage of this approach is the higher data acquisition rate - 20 times faster than with the scanning mirrors. However, the time taken in post-processing mitigates this, and there is no clear advantage over the other approaches for crack closure or low-cycle fatigue studies.

A minor disadvantage is the physical size of the circuit board on which the linear array detectors are mounted. As the spacing between indentations decreases, the fringe spacing increases, and one must move closer to the specimen to capture three fringes. However, the detector itself could be mounted off the board and connected by a cable. This would slow down the maximum scan rate which is much higher than the one actually used; therefore, it should cause no problems.

TYPICAL RESULTS AND DISCUSSION

A typical load-displacement plot for small cracks on the edge of a 2.3mm thick specimen of 2024-T3 aluminium is shown in Figure 14.5. The indents were initially placed at the centre of the 210μm surface crack. As the specimen was cyclically loaded, the crack grew, and another record was taken when it was 1.23mm long. The specimen was loaded equally between tension and compression, and the zero displacement arbitrarily taken at the most negative load. There is some hysteresis, especially for the smaller crack; this is typical. Note the linear region at the higher loads for the longer crack. The figure also demonstrates a usual way of determining P_{op}; straight lines are drawn through the upper and lower linear regions, and their intersection defines the opening load.

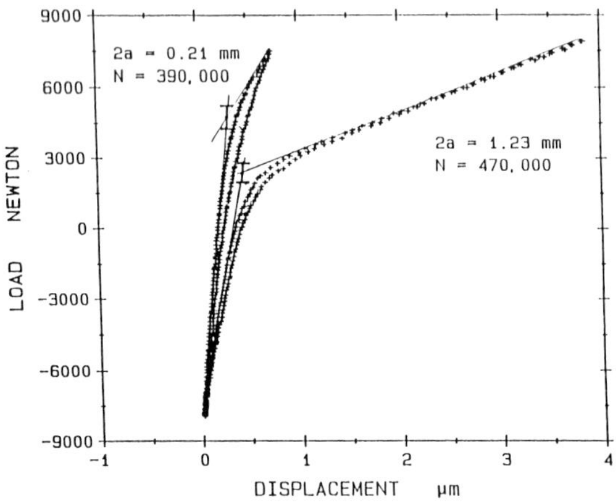

Figure 14.5: Typical load-displacement records for fully reversed fatigue loading taken with the incremental-sweep system. The two records were taken with the same set of indents in the centre of a surface crack, but for different lengths of the crack. A typical procedure of fitting straight line to the linear portions of the record and using their intersection to define the opening load is shown

Figure 14.6 shows a load-displacement record from indents placed 50μm behind the tip of an edge crack in an IN-100 specimen at 650°C. There is very little displacement here because the maximum load was purposely kept low to avoid crack growth as the specimen was loaded through 50 cycles. Again, there is hysteresis on the first cycle, but the fiftieth cycle shows none. Measurements at these high temperatures are possible because the material remains reflective; it is a nickel-based superalloy used in gas turbines.

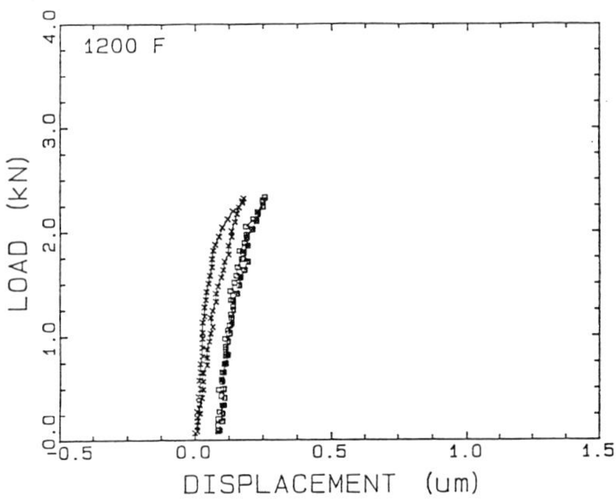

Figure 14.6: *Load-displacement records taken 50 µm behind the tip of an edge crack at 650C on a nickel-based superalloy with the single-sweep system. The crosses represent the first cycle, and the squares the fiftieth cycle*

Typical linear array detector system results are shown in Figure 14.7. Displacements were measured at a position 80 µm behind the tip of a crack in a compact tension aluminium specimen. Note the large number of data points and the very pronounced linear region at the higher loads. The primary use of this particular version of the ISDG has been to use the measured compliances of the upper linear region to determine the crack length. A routine for fitting a straight line to the upper portion is included in the software.

One can see that all three systems have essentially the same resolution - of the order of 0.01µm. The speeds of data acquisition are also basically the same if recording on a cycle-by-cycle basis. Loading for high-cycle fatigue studies is usually done at frequencies of the order of 20hz or faster, so neither approach is fast enough to record to display load-displacement near the crack tip. However, the linear array detector system is 20 times faster in actual data acquisition. If one were running a high-cycle fatigue test and wanted to record load-displacement every 1000 cycles (say), then the linear array detector would be preferable. A pair of linear array detectors is also cheaper than a pair of scanning mirrors with controllers. On the other hand, there is a fair amount of custom-made circuitry associated with the linear array system described in [6]. Choosing a system is therefore somewhat a matter of taste.

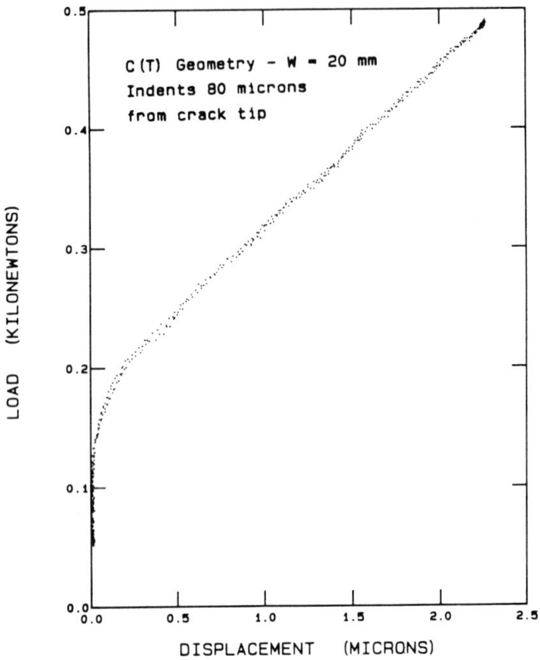

Figure 14.7: *A load-displacement record using the linear array detector. The indents were 80 μm behind the tip of a long crack in a compact tension specimen*

It is nearly always desireable to make a measurement system faster. The ISDG has been used quite a bit for impact studies with the fringe motion recorded on photomultiplier tubes, but the resolution is limited to 1/2 fringe. Higher speed solid-state detectors will surely be forthcoming which will improve the linear array detector approach. An analogue approach in which a rotating mirror sweeps the fringes patterns past photo-multiplier tubes followed by phase-measuring circuits offers an intriguing possibility.

Crack-opening displacement of small cracks have been measured by other methods, e.g. with the aid of a scanning electron micro-scope [8], but this is a cumbersome procedure that produces few data points. Small cracks require a small transducer, and the tiny indents of the ISDG serve that purpose. Also, the three systems described record on a realtime basis and store the data for easy later analysis.

ACKNOWLEDGEMENTS

The support of the National Science Foundation, NASA, and the Air Force Wright Aeronautical Laboratories over a period of several years is gratefully acknowledged. Technical discussions with Dr. T. Nicholas of Wright Aeronautical Laboratories have been especially informative and encouraging. The earliest version of the minicomputer-controlled ISDG was developed with significant contributions by graduate students M. Ward and M. Guillot. M. Karweit of Hopkins made valuable suggestions about the single-sweep approach.

REFERENCES

[1] Ritchie, R.O. and Lankford, J., Editors,"Small Fatigue Cracks" The Metallurgical Society, Warrendale, PA (1986)
[2] Jenkins, F.A. and White, H.E., "Fundamentals of Optics", McGraw-Hill, New York (1957)
[3] Sharpe, W.N., Jr.,"Applications of the Interferometric Strain/Displacement Gage", Optical Engineering, 21, 3, (1982) pp. 483-488
[4] Guillot, M.W. and Sharpe, W.N. Jr., "A Technique for Cyclic Plastic Notch Strain Measurement", Experimental Mechanics, 23, (1983) pp. 354-360
[5] Sharpe, W.N., Jr., "A System for Crack-Opening-Displacement Measurement and Photomicrography of Cracks at High Temperatures", Proceedings of the Eighth International Conference on Experimental Mechanics, Amsterdam (1986)
[6] Hartman, G. and Nicholas, T., "An Enhanced Laser Interferometer for Precise Displacement Measurements", Experimental Techniques, (February 1987), pp. 24-26
[7] Nicholas, T., private communication
[8] Morris, W.L. and Buck, O., "Crack Closure Load Measurement for Microcracks Developed During the Fatigue of Al 2219-T851", Metallurgical Transactions A, 8A, (1977) pp. 597-601

15 Recent advances in instrumentation for experimental stress analysis using thermoelastic techniques

D.J.Berry, G.M.Everett & R.H.Owens

SUMMARY

The thermoelastic effect has been employed to provide non-contacting, full-field stress mapping facilities for the stress engineer. The instrumentation making use of the thermoelastic principle is now well established for both experimental studies and for integration with the validation of analytical methods.

With the increasingly widespread use of the thermoelastic effect to measure dynamic stresses, so the theoretical and practical considerations governing its use have been refined. This chapter traces the historical background to the development of the technique and sets out the basic theory. Recent findings that challenge certain aspects of the basic theory are discussed. The instrumentation making use of the thermoelastic effect is outlined and the application of this instrumentation is discussed.

NOTATION

T absolute temperature
C_ε specific heat at constant stress
C_p specific heat at constant pressure
Q heat input
ρ mass density
σ_{ij} stress change tensor
ε_{ij} strain change tensor
K_m thermoelastic material constant ($\frac{\alpha}{\rho C_p}$)
α linear thermal coefficient of expansion
$\Delta\sigma$ change in the sum of the principal stresses (σ_1, σ_2, σ_3)
E Young's Modulus
ν Poisson's Ratio
ΔT reversible surface temperature change

HISTORICAL BACKGROUND

The phenomenon known as the thermoelastic effect is the change of temperature that accompanies the adiabatic elastic deformation of a body. Weber [1] (1830) observed this effect in a vibrating wire.

Lord Kelvin developed a theory of the reversible thermoelastic effect in 1853 [2] which applies to homogeneous isotropic materials under adiabatic loading conditions. This theory predicted a linear relationship between the change in the sum of the principal stresses and the temperature changes produced. The absolute accuracy of Lord Kelvin's formula has been verified a number of times since. Compton and Webster [3] (1915) used highly sensitive contacting resistance thermometers to verify the formula to an average accuracy of 0.07%.

Advances in infra-red technology enabled Belgen [5,6] (1967) to conduct the first non-contacting observations of the thermoelastic effect. He also investigated the effect of frequency of load cycling, specimen geometry and stress distribution on the attenuation of the thermoelastic effect by heat transfer. Adiabatic conditions were found to exist for a wide range of practical test conditions.

Anticipating further advances in remote temperature (radiation) sensing techniques, experimental systems were built and the early results reported by Mountain and Webber [7]. Worldwide patents were applied for, and granted from 1978, to cover the design of stress measuring systems based on the thermoelastic effect. These systems are manufactured and marketed by Ometron Ltd. under the name of SPATE 8000 (SPATE is derived from Stress Pattern Analysis by measurement of Thermal Emission).

Figure 15.1: SPATE 8000 Stress Analyser

BASIC THEORY

The basic thermoelastic relationship for the temperature change,

ΔT, caused by a change in the stress state in a linear, homogeneous, elastic solid was derived by Kelvin from the laws of thermodynamics in the form

$$\Delta T = \frac{T}{\rho C_\varepsilon} \frac{\delta \sigma_{ij}}{\delta T} - \varepsilon_{ij} + \frac{Q}{\rho C_\varepsilon} \qquad (15.1)$$

where

T is the absolute temperature
C_ε is the specific heat at constant strain
Q is the heat input
ρ is the mass density
σ_{ij} is the stress change tensor and
ε_{ij} is the strain change tensor.

The partial derivations in Equation (15.1) are readily obtained from the stress-strain equations for an isotropic elastic solid, and for adiabatic conditions in which $Q = 0$ the equation reduces to

$$\Delta T = \frac{-E \alpha T}{\rho C_\varepsilon (1-2\nu)} \varepsilon_{ii} \qquad (15.2)$$

in which ε_{ii} represents the sum of the three normal strain changes. Using the established relationship between C_ε and C_p, the specific heat at constant pressure, and eliminating the normal strains in terms of the normal stresses, Equation (15.2) becomes

$$\Delta T = -\frac{\alpha}{\rho C_p} T \sigma_{ii} \qquad (15.3)$$

Recognising ε_{ii} as the first invariant, Equation (15.3) may be written in the form:

$$\Delta T = -K_m T (\sigma_1 + \sigma_2 + \sigma_3) \qquad (15.4)$$

where K_m, the thermoelastic constant, equals $\alpha / \rho C_p$ and σ_1, σ_2 and σ_3 are the changes in the principal stresses.

More simply $\Delta T = - K_m T \Delta\sigma$

This general relationship enables the user to obtain quantitative stress data via a theoretical or known stress calibration procedure. The latter, which makes use of known stress data for the material under test is the more widely used in practice, but the theoretical method can be employed where the physical constants are known.

This general relationship has recently been challenged by Machin, Sparrow and Stimson [4]. Precise measurements were carried out over a range of mean static stresses onto which a cyclic load was superimposed. The findings of this work suggest that the thermoelastic constant K_m, in fact varies with mean stress.

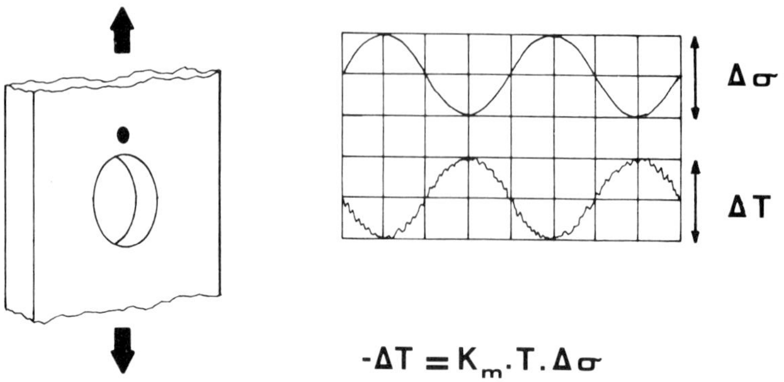

Figure 15.2: Thermoelastic theory for cyclically loaded structures

The practical work has, very recently [7] been complemented by a theoretical study which argues the stress dependence of the thermoelastic 'parameter'. The authors conclude that whereas the stress dependence of the thermoelastic 'parameter' does not materially affect the practicality of obtaining good quantitative stress data, especially where the known stress method of calibration is used, it does open up the possibility of using the technique to determine the actual level of the mean stress, i.e. it offers potential for the assessment of residual stresses.

THEORETICAL STUDIES

The basic theory of thermoelasticity has been developed by a number of workers [8, 9, 10, 11, 12]. Cummings, Harwood and Pukas of the National Engineering Laboratory have extended the understanding of factors influencing calibration and have developed the instrumentation to incorporate the facility for handling random excitation of the structures. Fundamental studies at the University of Manchester have included work on fracture mechanics and fatigue life assessment. For the latter the relationship between the stress intensity factors K_I and K_{II} and the SPATE (Stress Pattern Analysis by Thermal Emission) data is described for a wide plate containing a central crack. Stanley and Chan derive K_I and K_{II} using the Westergaard equations for the elastic stresses in the vicinity of the crack tip. The stress intensity factors, which are of critical importance in the assessment of the fatigue life of a cracked component, can be related to fatigue crack growth rate by the Paris Law

$$\frac{da}{dN} = C(\Delta K)^m$$

where
a is the crack length

N is the number of cycles
K is the stress intensity factor range
C and m are material constants
using SPATE data, da/dN can be derived experimentally using the full field stress mapping facilities of the equipment to produce stress maps, contours will become skewed about the line of the crack as the crack develops. The experimental data obtained for mode I loading were found to be in close agreement with numerical values and further work is planned for mode II and mixed mode loading.

Studies at the Royal Aircraft Establishment, Farnborough have demonstrated the relationships between SPATE 8000 output and the strains in multidirectional laminated fibre composites. The general principles for thermoelastic stress analysis of a unidirectional composite state that the adiabatic temperature change T at any point on the surface of an orthotropic material is given by

$$\Delta T = -K_m(\alpha_j \Delta\sigma_j + \alpha_k \Delta\sigma_k)$$

where

α_j, α_k = coefficients of thermal expansion in the principal stress directions

$\Delta\sigma_j$, $\Delta\sigma_k$ = change in principal stresses

The above applies to a unidirectional fibre composite laminate on the assumption that it behaves as a homogeneous material. However, more generalised conditions can be accommodated by representing α and σ by vectors for the principal stress and material axes.

From the equations developed by this approach it can be shown that

$$S = K(\Delta_{\sigma'} + \alpha\Delta\sigma_2) \qquad \text{where} \qquad K = \frac{AT\alpha_1}{\rho C_p}$$

where S is the SPATE 8000 output and A is a constant, determined experimentally, to accommodate the effects of surface emissivity and the fact that the process may not be perfectly adiabatic.

The work demonstrates the need to provide information relating to the lay-up of the composite and the characteristics of the fibre/epoxy material to permit quantitative interpretation of thermoelastic data.

DETECTION OF CYCLIC TEMPERATURE CHANGE

The SPATE 8000 system detects the infra-red flux emitted from points on an observable surface as a result of the minute cyclic temperature changes in a cyclically stressed structure or component. The SPATE 8000 Stress Analyser works over a cyclic stress frequency range of 0.5Hz to 20,000Hz. This enables assessment of most structures under realistic service conditions.

SPATE 8000 can measure instantly at a single point on the structure under test, produce a linescan of typically 120 data points in 2 minutes or less, or produce full field stress maps in a few minutes. Additional single point and lines of data can be recalled from scans stored in the system's non-volatile memory.

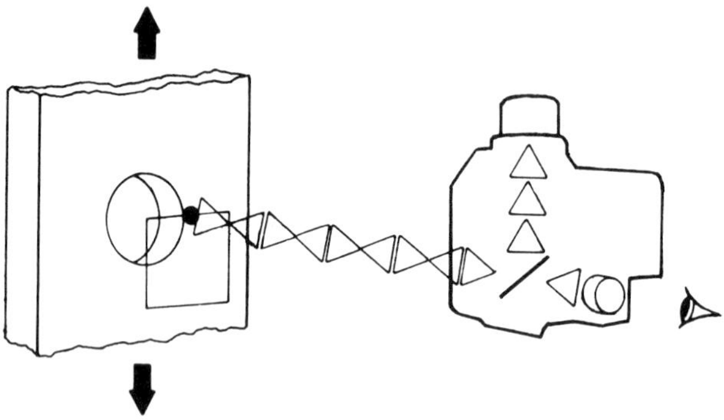

Figure 15.3: Visual identification of measurement point

The size of the measurement spot, effective gauge size, increases with distance from the structure under test. With the SPATE 8000 scan unit 0.3m from the structure the measurement spot is only 0.5m diameter, allowing fine stress detail to be measured. With 1.0m between the scan unit and the structure the measurement spot is 1.2mm diameter. The scan unit contains a high performance infra-red detection system, cooled by liquid nitrogen. It is the performance of this detector which gives the excellent sensitivity to cyclic temperature changes of 0.001°K resulting from the thermoelastic effect. The infra-red detection system works over a range of test structure surface temperatures. Surface temperatures of over 600°C have been successfully stress mapped. Two computer driven mirrors are located in the scan unit to produce a raster type of scan. The scan line, or area, can be operator selected anywhere within a 25° field of view. The limits of the area or line to be measured are fixed using a handset attached to the scan unit, and the projected white light spot. The number of measurement points, data acquisition time and display mode are all entered through a fixed keyboard following a 'user friendly' set of computer prompts.

The detected infra-red signals are correlated in frequency, magnitude and phase with a load derived reference signal. The frequency correlation ensures that measurements are made at the load signal frequency only, thereby eliminating environmental effects; the magnitude correlation gives an output proportional to $\Delta\sigma$, and the phase correlation determines the sign of $\Delta\sigma$. This method of correlation also provides a real time stress value which allows the stress engineer to measure a large number of discretely identified points very quickly, like an 'optical stress gauge'.

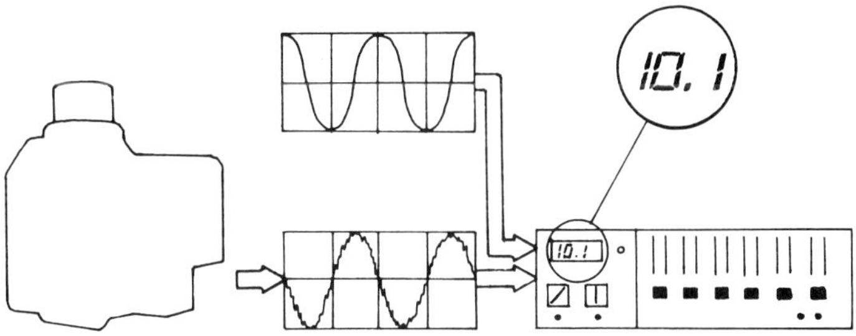

Figure 15.4: *Correlation between temperature and reference signals*

APPLICATIONS

The range of applications of SPATE 8000 extends to virtually all engineering materials and from small components, such as turbine blades, through to massive civil engineering structures.

The work described so far has been in relation to simple test specimens being loaded at uniform frequency and amplitude. The SPATE 8000 system can also be used on complex structures and under complex loading conditions. For complex shapes, the non-contacting nature of the instrument means that, provided a line of sight can be established between the test structure and the SPATE 8000 scan unit, stress data can be obtained. There is no need to view a structure normal to the surface, as in photo-elastic techniques for example, and experimental work has established that there is no significant attenuation of the signal at 70°F from normal to the surface. Similarly, because of the long wavelength infra-red radiation, in the range 8-14 μm, a much greater depth of focus is obtained than at visible wavelengths. These factors together mean that complex shapes, curved surfaces and assemblies can be analysed.

Complex loading can be accommodated directly by the SPATE 8000 system. Running engines, for example, can be used to generate a cyclic load. A narrow band pass filter is then used to select specific frequencies for the load derived reference signal and to determine the stress distributions associated with those frequencies of dynamic load.

The above technique has been found to be particularly useful for determining 'dangerous frequencies' for structures. A typical example of work using a running engine to generate dynamic loading is shown in Figure 15.5 with the corresponding stress map shown in Figure 15.6.

Here the engine of a fork lift truck was run at various speeds and specific frequencies selected to determine the stress distribution around a headlamp bracket. In this real situation in which a manufacturer had experienced service failure, the SPATE 8000 system revealed the highly localised nature of the stress distributions (so localised that it had been missed using strain gauges) and was able to quantify the magnitude of these stresses.

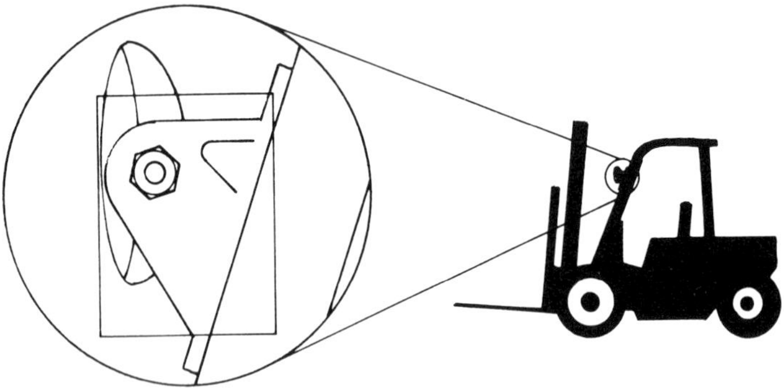

Figure 15.5: Schematic showing area scanned on running fork lift truck

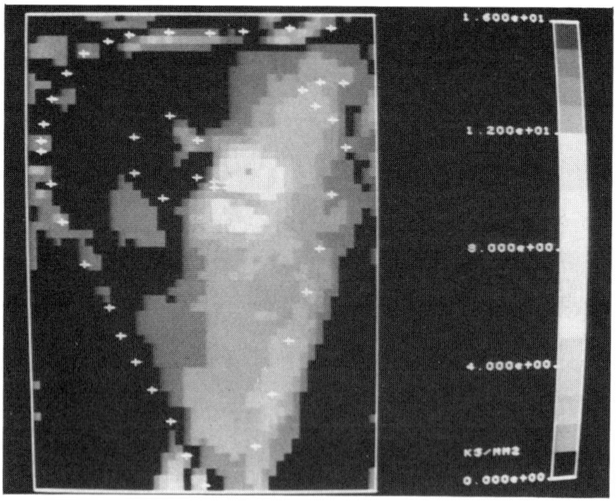

Figure 15.6: SPATE 8000 stress map from fork lift truck

It is interesting to note that because the SPATE 8000 technique is both non-contacting and rapid, it is possible to effect practical modifications to a component such as this bracket and to quickly determine the effectiveness of the remedial action. Similarly stress maps of structures and componments such as turbine blades and anchor chain links (Figures 15.7 and 15.8) can be produced with equal facility.

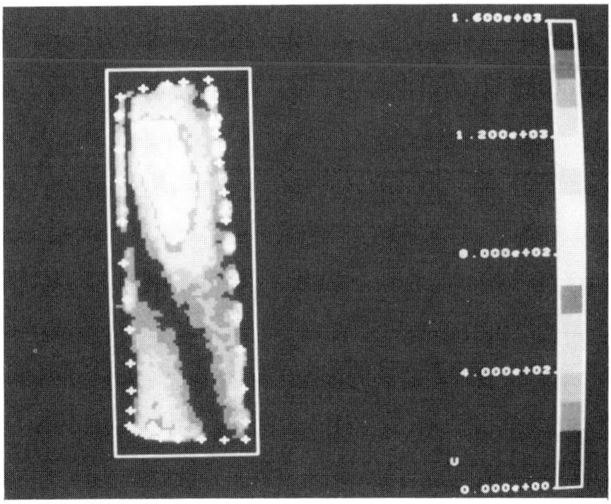

Figure 15.7: *SPATE 8000 stress map of a compressor blade excited in resonance (1st torsional mode)*

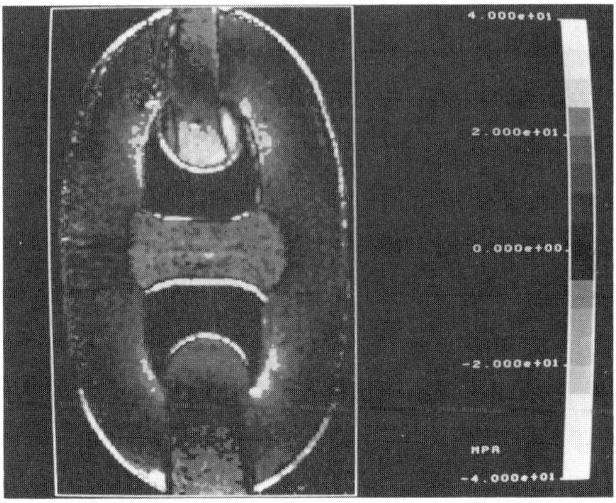

Figure 15.8: *SPATE 8000 stress map from chain link dynamically loaded in tension (courtesy National Engineering Laboratory and Bruce Anchors Ltd.)*

COMPARISON BETWEEN FULL FIELD EXPERIMENTAL METHODS

Photoelasticity

Transmission (i.e. model) technique

This technique requires the manufacture of a scale model in transparent plastic, which then has a beam of plane polarised light passed through it. Under a static load an interference fringe pattern (i.e. isochromatics) will be visible in the model when viewed in a polariscope. The fringes must be counted from a datum to calculate the principal stress difference (i.e. proportional to the maximum shear stress) at a point in the model. This stress may then be related to the corresponding stress in the actual component by a simple formula. By means of rotating the polariscope, taking further measurements (i.e. isoclinic evaluation), and going through a very laborious procedure the magnitude and directions of each principal stress can be determined over the entire surface.

The potential ability to separate the principal stresses is, of course, very useful for failure prediction. However, for the great majority of two dimensional models the highest stress will occur at a free surface, i.e. there will be only one principal stress. Stress analysts are normally primarily interested in determining the magnitude and position of the highest stress rather than obtaining a complete description of the entire stress field in a component. Therefore, the principal stress separation technique is not always performed, particularly if the component is manufactured from a ductile material which is known to obey the maximum shear stress failure criterion.

A stress freezing technique, in which the loaded model must be heated (135°C) until the secondary bonds break down, can be used to retain the stress pattern within the model. Cooling the model and then sectioning it will allow a three dimensional stress analysis to be undertaken.

The fringe patterns need skilled interpretation since it is not immediately apparent where the zero datum is located, and whether stresses are tensile or compressive. The model must, of course, be a very accurate representation of the actual component, and thus great care must be taken in its construction. Implicit in the modelling technique which, of necessity has to make use of a plastic rather than the actual structural material, is the fact that many practical situations such as anisotropy, inhomogeneity or components with individually moving parts cannot be modelled. Moreover, a precise knowledge of the loading conditions is necessary, and this is not always easy to evaluate or simulate.

Reflection (i.e. photoelastic coating) technique

This technique is useful for surface stress measurement since there is no necessity to manufacture a scale model. A birefringent (e.g. double refracting) plastic coating is bonded to the structure with a reflective adhesive. The polarised light must, therefore, pass through the coating twice. The procedure for producing a uniform coating is rather laborious, time-consuming and messy. Moreover, only about one square foot (approx. 300m x 300mm) at a time can be covered. For larger structures the sections must be cut to shape and bonded together. Results can only be obtained in areas accessible to light, and care must be taken to ensure that the coating does not reinforce the component. This technique is limited in field applications by the high level of skill and judgement required from the technician applying the coating. It is subject to errors due to ambient temperatures and humidity fluctuations and other environmental effects, and suffers from the fact that the very bonding of a coating may stiffen or modify the structure and hence seriously affect the validity of the results obtained. The maximum sensitivity which can be expected from photoelasticity is $\pm 10\mu\varepsilon$.

Moiré interferometry

This is essentially a laboratory rather than a field technique more applicable to obtaining initial design information from model analysis rather than to the testing of engineering structures. If a diffraction grating is bonded to the surface of a component, a laser-interferometer containing a reference grid will reveal fringe patterns which are contours of constant displacement in a particular direction when the component is subjected to a load. The technique can only be used on a flat surface, and has an accuracy of $\pm 50\mu\varepsilon$ when a very high density grating (i.e. $>10^6$ holes/mm^2) is used; the accuracy is related to the hole density of the grating. The rather poor sensitivity of the Moiré technique means that it is more suited to plasticity investigations rather than the relatively low strain range normally encountered in experimental stress analysis. Moreover, the fact that the fringes are in the form of displacement contours rather than stress is a great disadvantage for stress analysis since there is no automated procedure available as yet for the conversion of the Moiré fringe patterns to stress. There is also a manufacturing problem in producing the very high density gratings required for reasonably sensitive measurements. These gratings are therefore rather expensive. The Moire technique has, however, been used to track damage growth in composite materials, to evaluate fracture mechanics parameters, and to indicate the non-uniform nature of residual stress fields. Research is continuing to improve the accuracy and convenience of the technique but it again suffers from the basic disadvantages that it requires a grating to be bonded to the structure and hence may cause modification to that structure. It is only applicable to static loading.

Laser holography

This technique consists of recording photographically the diffraction pattern produced by laser illumination of a component, and later reconstructing its image from the diffraction pattern. The image has depth but is not a complete 3-D description since it only reveals the parts of the component which were illuminated by the laser. The basic diffraction pattern is produced by recording the interference pattern between the laser beam and the light back-scattered from the component. A pulsed laser can be used for dynamic applications. The hologram stores information as to the intensity of the light back-scattered from the component and its phase relative to the laser beam. The phase information is related to the position of the component, and it is this which provides the three dimensional property of the reconstructed holographic image.

Like the Moiré technique the holographic pattern is in the form of displacement rather than stress contours, but it is insensitive to displacements which are other than out-of-plane. The technique requires complex and expensive equipment which must be set up exactly in order to give accurate data. Under ideal conditions the theoretical accuracy will be better than 1μm (i.e. similar to the wavelength of the laser). However, practical constraints mean that this high accuracy is unlikely to be achieved.

It should be noted that Michelson type interferometer option is now available for the SPATE 8000 system. This interferometer developed originally by UKAEA [13], can be interfaced directly to SPATE 8000 thereby providing full field vibration/displacement data from the same experimental set up as for stress analysis. In this system, called a Vibration Pattern Imager (VPI), a He-Ne laser is used so that signals can be obtained from poorly reflecting surfaces with response to frequencies approaching 1MHz. As the surface moves the output of the interferometer fluctuates periodically, with one cycle corresponding to a movement of $\lambda/2$ (0.3164 μm). The frequency of the signal is thus proportional to this frequency or velocity. Combined with the SPATE 8000 software these signals can be used to give full field, non-contacting measurement of vibrating surfaces to a high spatial resolution and with the facility to internally calibrate directly in terms of the wavelength of the laser. Additionally, the system is not limited to small vibration amplitudes and does not require a costly specialised, controlled environment for operation.

Brittle lacquer

This technique involves the spraying of a suitable brittle coating on to the surface of a component. The brittle coating will crack perpendicular to contours of maximum principal stress if the maximum principal strain exceeds the threshold value of the lacquer. The threshold strain is about 500με but depends on temperature and humidity, thereby making reproducibility difficult to achieve. It is a cheap and useful on-site technique if maximum tensile stresses are of primary interest, but it is imprecise, labour intensive, and is therefore normally used as a preliminary to other more precise techniques. Lacquers can be somewhat toxic and are broadly insensitive to compressive stresses.

Thermoelasticity

Stress pattern analysis by thermal emission is a new technological development which is based on a hitherto unexploited physical phenomenon known as the thermoelastic effect. The thermoelastic effect refers to the thermodynamic relationship between changing stress in a component and the temperature changes in the component produced by those stress changes. Under adiabatic conditions, produced by dynamic loading, the change in temperature at a point in a component is proportional to the change in the sum of the principal stresses at that point.

The advantages of the thermoelastic technique as embodied in SPATE 8000 are:

(a) A full field stress pattern can be obtained very rapidly.
(b) Only minimal surface preparation is required (i.e. cleaning and spraying with high emissivity paint).
(c) Complex geometries can be evaluated almost as easily as flat surfaces since the thermoelastic response is not particularly sensitive to viewing angle, and the instrument has a wide depth of focus.
(d) A spatial resolution of 0.5mm can be achieved.
(e) The sensitivity of the instrument is higher than can usually be achieved using other full field techniques. The stress sensitivity is dependent on the component material, but 0.4MPa (60 lbf/in^2) can be attained in aluminium. Furthermore, the technique can differentiate between compressive and tensile stresses.
(f) The technique is unique in being a non-contacting stress measurement on a real structure.
(g) The field of view is altered simply by moving the camera and/or the limits of the scanning mirrors. Similarly the number of stress measurements within the field of view is preselectable by the operator.
(h) A colour coded stress pattern, displayed directly on a VDU, is simple to interpret.
(i) The stress patterns may be archived on a dedicated floppy disk unit or transferred to cartridge disk on a mainframe for detailed examination or post-processing.
(j) High temperature measurements of qualitative stress patterns may be undertaken.
(k) The standard system is capable in principle of extension to dynamic service loading applications.
(l) Results correlate well with both theoretical and other experimental methods and the system integrates readily to existing testing practice.
(m) In addition to full field data, individual point or line data can be produced with a consequent saving in data acquisition time.
(n) Experimental methods are available to accommodate displacements normal to the viewing direction (which would otherwise smooth the measurement stress data over the movement range).
(o) The technique works on virtually all structural materials including metals, composites, plastics and ceramics.
(p) Design or manufacturing features, such as welds, fasteners

etc., can be assessed directly.

(q) No controlled environment is necessary for operation.

The disadvantages of SPATE 8000 are listed below:

i) The system is relatively expensive - although the actual cost of producing stress data is small.

ii) For accurate quantitative work the loading must be dynamic at a frequency greater than 3Hz for steels in order to achieve adiabatic conditions. The stresses must also remain within the elastic range of the material (isentropic conditions). However, good qualitative data can be obtained even when these conditions are not met.

iii) The stress pattern is in the form of principal stress sum. Individual principal stresses and directions cannot be separated. Pure shear produces no thermoelastic output.

iv) Only surface stress can be measured.

v) Optical access is required (special infra-red mirrors can readily extend the field of access).

vi) Calibration from material properties requires a precise knowledge of physical data which may be difficult to obtain for the component material. However, a simple experimental 'known stress' method is available whereby a strain gauge rosette is used to provide calibration data.

vii) The system measures dynamic stresses. However recent work has suggested the potential of the thermoelastic effect for the assessment of residual stresses.

CONCLUSIONS

The thermoelastic effect has been successfully used as the basis for instrumentation providing full field stress data. This instrumentation complements and offers significant advantages over other full field methods. Furthermore, thermoelastic measurements have been shown to correlate well with strain gauges whilst at the same time yielding benefits of full field data which can be used to provide more comprehensive descriptions of the structure under test.

Fracture mechanics, fatigue studies, crack propagation analysis have been undertaken and the underlying theory determined experimentally using SPATE 8000. Similarly, the instrumentation has been applied to a wide range of materials and geometries including weldments, plastic models and composites. Calibration of data to provide quantitative results has been considered for non-homogeneous materials. Additionally, the effect of mean stress on the thermoelastic parameter has been studied. This latter work has yielded the potential possibility of extending the application of thermoelasticity to study not only dynamic stresses but also residual stress. The addition of full field vibration pattern imaging from the same experimental set-up as for stress analysis further extends the range of applications of the instrumentation.

REFERENCES

[1] Weber, W., Pogg Ann der Phys 20, 177 (1830)
[2] Thomson, W. (Lord Kelvin), Jnl Maths (1855), reprinted in Phil Mag 5, pp. 4-27 (1978)
[3] Compton, K.T. and Webster, D.B., Phys Review 5, 159 (1915)
[4] Machin, A.S., Sparrow, J.G., Stimson, M.G., Brit Soc for Strain Measurement 23, 1, pp. 27-31 (1987)
[5] Belgen, M.H., ISA Trans 6, pp. 45-53 (1967)
[6] Belgen, M.H., NASA Report CR-1067
[7] Wong, A.K., Jones, R., Sparrow, J.G., J Phys & Chem of Solids (1987)
[8] Cummings, W.M., Harwood, N., Proc SPIE 590, pp. 268-276 (1986)
[9] Stanley, P., Chan, W.K., Proc 2nd Int Conf on Stress Analysis by Thermoelastic Techniques, SPIE (1987)
[10] Potter, R.T., Proc 2nd Int Conf on Stress Analysis by Thermo-elastic Techniques, SPIE (1987)
[11] Harwood, N., Cummings, W.M., Proc 2nd Int Conf on Stress Analysis by Thermoelastic Techniques, SPIE (1987)
[12] Pukas, S.R., Proc 2nd Int Conf on Stress Analysis by Thermo-elastic Techniques, SPIE (1987)
[13] Drain, L., Moss, B., QT News, 28 Nov 1986

16 Methods for fringe order determination in photomechanics

J.D.Hovanesian & Y.H.Hung

SUMMARY

Misinterpretation of the fringe order and errors in determining its value is one of the greatest, if not the greatest, shortcoming in photomechanics. Three methods are proposed which either improve or automate existing procedures. The methods are: (1) Carrier fringe methods, (2) Coherent compensation, and (3) Colour load coding.

INTRODUCTION

Most photomechanic methods involve the formation of interference fringes which are interpreted to yield quantitative data. Unfortunately, the fringe "phase" is not directly displayed and may require considerable skill in the course of its interpretation. This task is even more complicated when the fringe order is a non-monotonic function with respect to spatial coordinates. When no convenient boundary conditions are in existence, the determination of the fringe order at a point may be difficult indeed.

The determination of the absolute fringe order, including its sign and fraction is paramount in photomechanics. Nobis and Vest [1] clearly indicated the importance of accuracy in determining the fringe order upon accuracy in displacement determination in hologram interferometry. In Moiré analysis, Voloshin et al [2] have shown that when the fringe orders are low in number, the need for accuracy in the fringe identification is even more critical. Post [3] developed fringe sharpening and multiplication techniques in order to facilitate fringe interpretation in photomechanics. Sciammarella and Rowlands [4] used numerical and analogue techniques to retrieve and process fringe information. Massie [5] used heterodyne interferometry and Sasaki and Okazaki [6] introduced a sinusoidally varying phase into the interferometer system in order to improve accuracy in surface profile measurement. In a separate paper [7] the same authors claim that accuracies of the order of 1nm are realizable. In addition to the above, many others have striven to improve fringe order determination.

Carrier fringe methods

In order to facilitate the determination of the fringe order on a full field basis, a linear fringe "carrier" system has been introduced which is coherently superimposed upon the fringe system owing to deformation, stress or strain. The linear fringe system dominates the phase variation due to deformation, hence the final fringe system is spatially quasi-linear and monotonic. As such, there is virtually no difficulty in determining the fringe order at any point with respect to any point of known deformation, stress or strain.

The carrier fringe system by itself may be characterized as:

$$I_c = I_o (1 + \cos kx) \qquad (16.1)$$

where

I_c = intensity of carrier fringes alone

I_o = object intensity without fringes

K = $\frac{2\pi}{p}$ or spatial wave number, where p is the spatial period

The above intensity expression assumes that the dominant linear carrier fringes are imposed in the x-direction. Because of the dominance of the carrier fringes, the addition of phase variations owing to deformation is small. The total fringe order is quasi-linear and monotonic. Without the carrier fringes, the intensity variation I_s owing to deformation, is denoted by:

$$I_s = I_o (1 + \cos\Delta) \qquad (16.2)$$

where I_s is the intensity of deformation fringes alone and Δ is the phase change due to deformation which may be expressed as $\Delta = \frac{2\pi}{\lambda} L(x,y)$; $L(x,y)$ is the optical path change.

Coherent or phase superposition of (16.1) and (16.2) yields

$$I = I_o [1 + \cos(\Delta + kx)] \qquad (16.3)$$

I = Intensity distribution for coherent superimposition of (16.1) and (16.2).

Equation (16.3) represents a modulated fringe pattern resulting from the phase superposition of the known carrier and the deformation. If the carrier dominates, the fringe pattern will have monotonically increasing or decreasing fringe orders, and thus allows the fringe order relative to any given point to be determined without ambiguity. The phase change due to deformation alone is obtained by subtracting the known carrier from the modulated fringe pattern. The process of data reduction is so straight forward that skillful human interpretation is no longer required.

In order for the resulting fringe pattern to have monotonicity, the gradient of the superimposed phase variation should not have a sign change. This condition is met provided:

$$|K| > \left| \frac{\partial \Delta}{\partial x} \right| \qquad (16.4)$$

The above equation implies that the spatial frequency of the carrier should be greater than that of the deformation for the carrier to dominate. This condition will be further explained in the following illustration.

As a demonstration of the method, a beam problem with known boundary conditions is shown in Figure 16.1a. The fringe system displayed in Figure 16.1b is obtained by standard hologram interferometry which depicts the beam deflection due to the loading. Figure 16.1c illustrates the correct displacement interpretation which was possible owing to the known loading and boundary conditions. Had Figure 16.1a not been present, ambiguities would have existed in the interpretation of the fringes. Note that there exists a "peak" and a "valley" in the deflection curve. Each of these reverses the sequence of the fringe orders. This slope change cannot be identified from the fringe pattern. In addition to the ambiguity of the fringe sequence, there is another ambiguity in determining the algebraic sign of the fringes.

The carrier method proposed overcomes both ambiguities (Figure 16.2b). With the introduction of the carrier fringes, the fringe orders increase monotonically giving no difficulty in determining the fringe order at each point and this is illustrated as the "Deformed Carrier Fringes (N_d)" (Figure 16.2). The crux of this method is the straightforward manner by which the fringe orders may be counted in Figure 16.2b compared to the ambiguity in Figure 16.1b. Figure 16.2c shows that the carrier fringes, N_o, varying linearly with respect to x, and the addition of the deformation fringes N_d produces a perturbed carrier fringes. The plot of the perturbed carrier fringe orders N_d does not have a reversal in the fringe sequence, i.e., no "peaks" and "valleys". This condition corresponds to

$$\frac{\partial N_d}{\partial x} \neq 0 \qquad (16.5)$$

which is equivalent to Equation (16.4). The above condition guarantees the absence of "peaks" and "valleys"; which ensures that the fringe orders of the perturbed carrier fringes will be monotonically increasing or decreasing and hence they can be determined without ambiguity. The fringe order of the phase change due to deformation is obtained by taking the difference between the linear carrier and the perturbed curve. The result is also displayed in Figure 16.2c. Both the fringe sequence and sign are unambiguously determined.

This method lends itself to automatic fringe counting by automated scanning, digitization or other computer controlled means of fast data acquisition.

Coherent compensation

Compensation methods have been employed in conventional photoelasticity for many years and some of these are described by

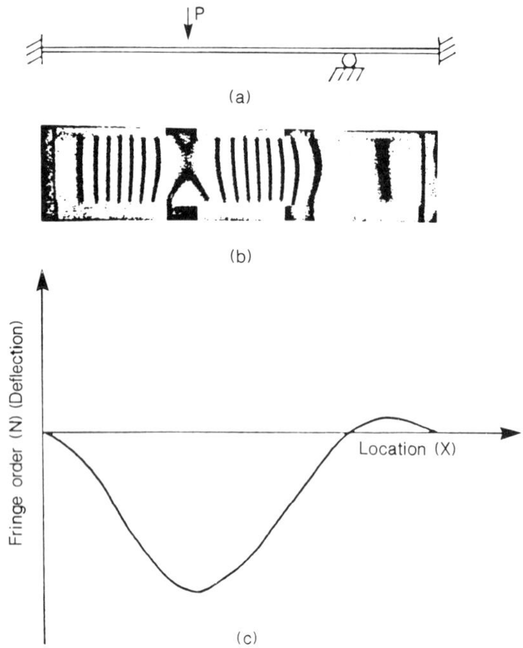

Figure 16.1: *(a) The loading situation*
(b) The corresponding conventional double exposure hologram
(c) The resulting deformation curve

Dally and Riley [8]. "Compensation" generally refers to techniques which are aimed at finding the fringe fraction together with the absolute order and sign identification. This is difficult in coherent interferometric methods such as hologram interferometry. Each fringe looks just like another. Real-time methods [9] are generally considered as excessively demanding and sandwich techniques [10], [11] may be time-consuming and require special and delicate apparatus.

The method to be described is particularly useful in hologram interferometry and its optical set-up is shown in Figure 16.3. An image hologram is made to localize the fringes on the hologram plate, although to image is not essential. During the first exposure, one reference beam is blocked and the other is used to make a hologram of the undeformed object. Next, a second exposure is made, this time of the deformed object using the other reference beam. Hence, each exposure is recorded with a different reference beam. During playback, both references illuminate the hologram simultaneously and produce high quality fringes similar to those obtained by the conventional double exposure technique.

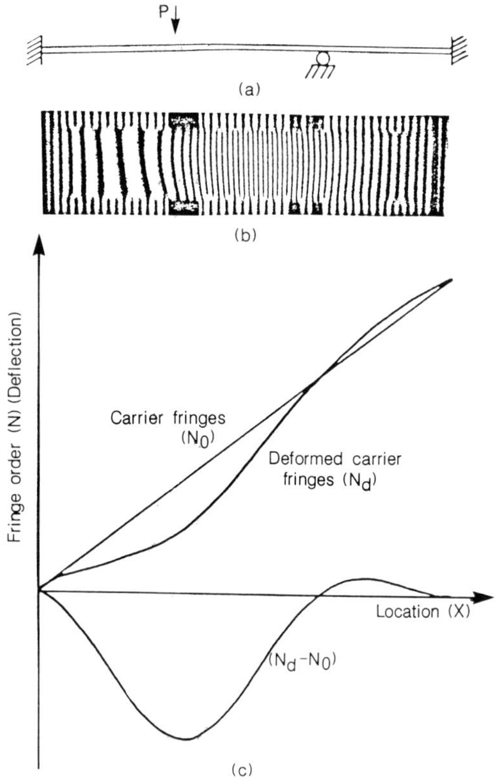

Figure 16.2: *(a) The loading situation; (b) Holographic interferogram with carrier fringes and deformation; (c) Plots of fringe order vs. location*

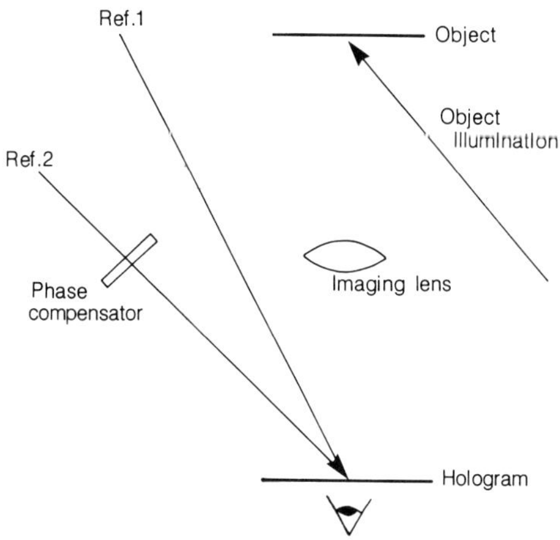

Figure 16.3: *Schematic of coherent compensation method*

In the above recording procedure, each of the two wavefronts corresponding to the deformed and undeformed object is coded with separate reference beams. This allows one of the wavefronts to be manipulated independently of the other. If a phase shift - that is, a change in the optical-path length - is imposed in one of the references, the fringes on the object will shift. This phase shift may be produced by a compensator which in this case was composed primarily of a thin plate of any transparent material with a refractive index close to but different from that of air (Figure 16.3). A Babinet-Soleil compensator is also quite suitable. The compensator is placed in one of the reference beams before the exposures are made. If a thin plate is used, rotation of the compensator plate changes the distance the light must travel through the plate and thereby the optical path length.

If a positive phase shift is imposed, all of the fringes will move uniformly in an algebraically increasing direction, i.e. fringe-order 1 will move towards order 2. The converse will occur if a negative phase shift is imposed. For this latter case, fringe-order 5 will initially move towards 4 and it can be shifted all the way to a known zero. This permits the absolute determination of integer fringes. Useful points such as local maximum/minimum and zeros can be identified. We will refer to these as peaks and valleys for the local maxima and minima respectively. If a positive phase shift is imposed, the fringes move toward and eventually disappear at the peaks; these are 'sinks' and are associated with local maxima. At valleys, a positive phase shift causes fringes to emerge; these are 'sources' and are associated with local minima. For fringe compensation, an added phase of Δ necessary to cause fringe number 4 to move to a point of interest allows for the absolute fringe order determination at this point. In this case, the fringe order at point A would be $4 + \Delta$.

In summary, this proposed method allows the unambiguous determination of the absolute fringe order including its algebraic sign and fraction. This compensation technique is a full-field method and requires only one hologram. While the information concerning the object's deformation is frozen, owing to the two exposures, the fringe shifting (compensation) may be performed in quasi real-time. This provides nearly all of the flexibility inherent in real-time hologram interferometry while allowing the advantage of playback at a later time. Concerning accuracy, this scheme lends itself to high-precision fringe-order determination. This is possible if the compensator device itself possesses an accurate calibration. Several commercial devices such as a Babinet-Soleil compensator are accurate to 1/1000 of a fringe, which in turn allows for the fringe-order determination to within ±1/1000 of a fringe.

This method also allows heterodyne interferometry to be performed in a frozen hologram. If a device is placed in one of the reconstruction beams which advances the phase of light passing through it, the output fringes will advance continuously. This time-varying phase may be monitored by a photo-detector and a phase meter.

Colour/load coding

This method is useful in identifying the fringe order while overcoming the "fade-out" phenomenon in classical white light interferometry.

In one of the two techniques herein described, two narrow band width sources, say λ_1 and λ_2 are used rather than broadband and continuous white light sources. These are illustrated in Figure 16.4. In this case, the initial condition corresponds to maximum light transmission as this is commonly the case in light field photoelasticity, hologram interferometry or in Moiré and speckle methods. Therefore, the zero order is an equal mixture of the two pure colours derived from the wavelengths λ_1 and λ_2. In Figure 16.4 this is designated as "M_0". The pure colours are designated as C_1 for λ_1 and C_2 for λ_2. As the orders increase, a characteristic sequence of coloured fringes occur for this combination of wavelengths. The initial, equal mix of C_1 and C_2 changes to an unequal mixture of the two until C_1 is totally extinct - then the pure colour C_2 appears. A sequence of pure colours and mixed colours continue until both colours extinct simultaneously and display "black". This occurs when the ratio of the number of extinctions of the individual wavelengths equals the inverse ratio of the wavelengths. As the orders increase beyond the null, eventually they become in phase again and both wavelengths simultaneously attain maximum intensity. At this point the equal mix of C_1 and C_2 again yield the mixture designated as M_0. From the lower equal mix of M_0 to the next higher M_0 is defined as a chromatic "fringe period".

The length and chromatic characteristics of a fringe period are noted as unique for a given pair of wavelengths. For narrow bandwidth spectral sources, an infinite number of fringe periods may be attained without degradation of fringe visibility. The sequence of the appearance of the pure colours and the mixes as well as their spacing reveals whether the order is increasing or decreasing.

Two methods have been demonstrated.

1. Two wavelength recording/playback method: This method is illustrated in Figure 16.5 as either holographic, another photomechanics recording/playback or a real-time set-up. In a holographic system, the lens shown is optional. In non-holographic systems, the reference beams would be deleted and the object illumination may be either by reflection or transmittance. In photoelasticity, the fringe patterns may be viewed in real-time. In a holographic system, the exposures of the two discrete sources may be recorded simultaneously or sequentially. In playback, however, both wavelengths must be used simultaneously. Whichever photomechanics and recording/playback methods are adopted, the use of two separate discrete wavelengths will display a fringe pattern like the one illustrated in Figure 16.4.

2. Two load method: The same result described for the two wavelength method may be attained by using two loads. An interferogram is recorded and "coded" at load P_2 with a single wavelength,

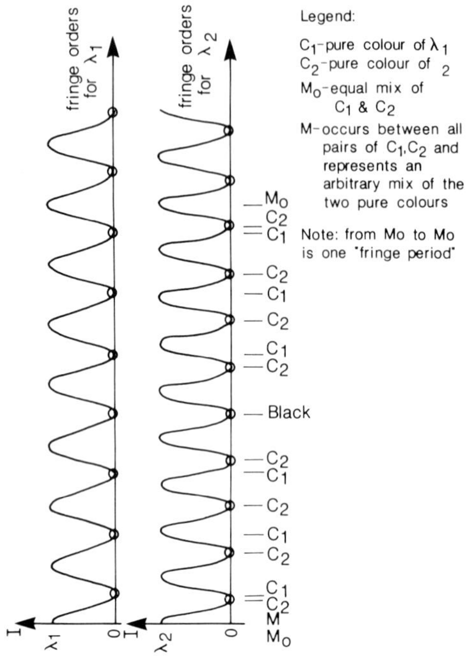

Figure 16.4: *Concept of the fringe period and sequence of colours in 2 wavelength fringe interpretation*

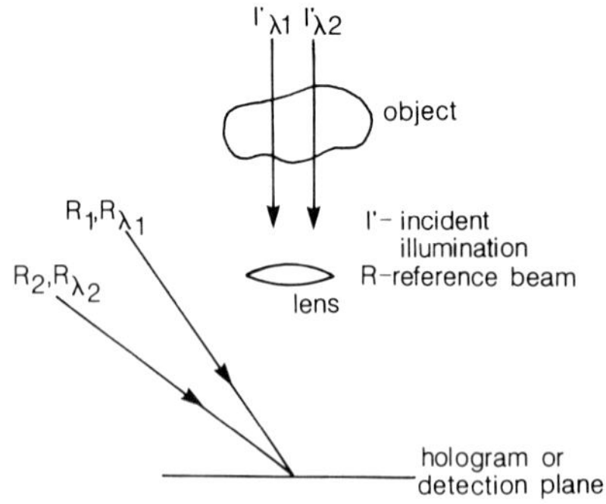

Figure 16.5: *Schematic of two wavelength recording/playback apparatus*

say λ_0 . The foregoing is followed by the recording and coding of an interferogram at another load P_2 with the same wavelength. Playback of the interferograms are performed with two wavelengths such that the fringe pattern associated with P_1 is displayed in colour C_2, associated with λ_1. Similarly, the fringes associated with the load P_2 are displayed in the colour C_2. The key is in coding during recording and decoding during playback.

In holography, this can simply be accomplished by the use of separate reference beams. During recording the reference R_1 in Figure 16.5 is used for the loading at level P_1 while R_2 is used for the P_2 loading. During playback, separate wavelengths are used for the two references simultaneously to yield a chromatic fringe pattern. In this case, a small angular and sphericity adjustment may be necessary in one of the references to ensure that the primary images generated from the two references will coincide. If image holograms are used, the latter requirement is relaxed.

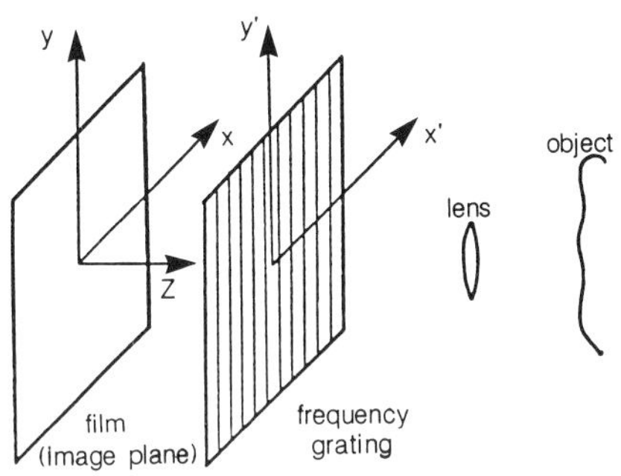

Figure 16.6: Coding scheme. Frequency grating superimposed on image

Another scheme of coding is illustrated in Figure 16.6 and it is particularly well suited to photoelasticity. The imaged plane (location of film) has a frequency grating placed close to it. The film will record the object's image with a superimposed linear grating. This is similar to the technique known as "carrier-frequency photography". Note that the illustrated carrier consists of either phase or amplitude lines parallel to the y-axis. The latter exposure (or double exposure) is recorded at a load P_1. The above process may be repeated for another load, say P_2, but the frquency grating would be rotated about the z-axis - say by 90°.

Decoded playback is illustrated in Figure 16.7. White light may be used, though it is not essential. The incident illuminating white light is adjusted to an angle α such that the first order diffraction yields the colour C_1 (associated with λ_1) and with its wave-normal parallel to the z-axis. This process yields a single colour fringe pattern associated with the load P_1. Simultaneously, another (not shown in Figure 16.4) incident illuminating beam may be provided to diffract the rotated carrier which was used in coding the fringes under load P_2. In the latter instance, the angle is adjusted to yield another colour say C_2 with its wave-normal also parallel to the z-axis. This playback method yields fringes identical to those illustrated in Figure

16.4.

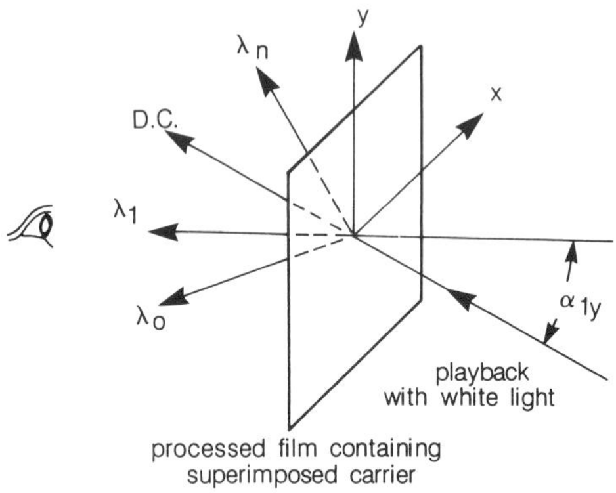

Figure 16.7: Decoding scheme

Fringe compensation, interpretation/desensitization:

In the methods described above for generating chromatic fringes, it is also possible to playback one colour at a time. This provides twice the fringe data which one would obtain from a single wavelength (or a single load) on a full-field basis. In generalizing the method to multiple, but discrete wavelengths, the amount of data may be increased accordingly. In a sense this is a form of compensation. Generalization to multiple wavelengths is discussed later in this chapter. Concerning the identification of the zero order, in the case of a dark-field system, the zero order fringe is black. The black fringe does however reappear at a higher order, precisely after one fringe period. In practical applications, this may not occur if the two wavelengths (or two loads) are only of slightly different magnitude. In this event, the inverse ratio of the integer extinctions would result in a fringe period containing a large number of orders.

It may be noted that the pure coloured fringes occur in ordered pairs with C_2 appearing first if $\lambda_2 > \lambda_2$. The spacing between these pairs increases with the magnitude of the order. This phenomenon allows for the determination of whether the fringes are increasing or decreasing. The relative spacing allows one to estimate the order itself.

Fringe desentilization is possible by selecting two wavelengths (or loads) which will result in the occurrence of several fringe periods. The incremental phase difference between two adjacent black fringes corresponds to the common multiple of λ_1 and λ_2. In this case, one would merely count the black fringes which represents a desentization.

The desensitization factor, i.e. the ratio of fringes owing to a single wavelength λ_1 divided into the number of black fringes is

$$\frac{\lambda_2 - \lambda_1}{\lambda_2}$$

In the two-load method, the desensitization factor is

$$\frac{\Delta P}{P_2} \text{ where } P_2 = P_1 + \Delta P$$

Generalization to multiple discrete wavelengths:

The two colour scheme is probably sufficient, but if one wishes more fringe data for compensation purposes, two wavelengths may be somewhat limiting. In this case, the procedures described above may be extended to display fringes in more colours. The previous section dealing with compensation is simply extended so that several single colour fringe patterns may be displayed separately to afford more effective compensation.

The problem of interpreting fringes and their fraction has heretofore represented an important limitation in photoholoelasticity. In classical photoelasticity, the use of conventional monochromatic sources has caused difficulty in determining peaks, valleys, saddle points and zeros. Continuous spectrally-distributed white light causes "fade-out" problems. The multiplexing of two or more discrete wavelengths can overcome the problems cited above. Furthermore, fringe desensitization is also possible and this technique should be most useful in problems associated with a high density of fringes. Such cases are frequently encountered in photoplasticity and in frozen stress techniques.

A major advantage of the methods proposed herein is the ability to code chromatic fringes without loss of visibility at higher orders. Furthermore, the ability to produce chromatic fringes from a single wavelength source and two loads adds additional flexibility. For example, one can predetermine the desensitization factor and at the same time eliminate the requirement of needing two or more lasers. The authors have recently employed digital image pseudocolouring to replace the laborious photographic recording and the subsequent optical processing. The work is planned to be reported in a future paper.

SUMMARY

Three methods: carrier fringe technique, coherent compensation and colour/loading coding have been presented in the chapter. These methods may be employed to aid in fringe order identification, compensation and interpolation in photomechanic techniques whose output are in the form of a fringe pattern, such as hologram interferometry, speckle interferometry, Moiré techniques, photoelasticity, shearography, etc. Of particular significance is the elimination of fringe ambiguities which has paved a way for automatic data deduction by digital image processing.

ACKNOWLEDGEMENT

This investigation was supported by the National Science Foundation (grant CEE-8219766). The support of Dr. Gifford Albright is apreciated.

REFERENCES

[1] Nobis, D. and Vest, C., "Statistical Analysis of Errors in Holographic Interferometry", Applied Optics, Vol. 17, (1978) pp. 2198-2204

[2] Voloshin, A.S., Burger, C.P., Rowlands, R.E. and Richard, T.G., "Fractional Moire Analysis Using Digital Imaging Techniques", Experimental Mechanics, Vol. 26, No. 3, (Sept. 1986) pp. 254-258

[3] Post, D., "Sharpening and Multiplication of Moiré Fringes", Experimental Mechanics, Vol. 22, No. 9, (1982) pp. 324-328

[4] Sciammarelli, C.A., Rowlands, R.E., "Numerical and Analog Techniques to Retrieve and Process Fringe Information", Proc. 5th Int. Cong. on Exp. Stress Analysis, Udine, Italy (1974)

[5] Massie, N.A., "Real-Tie Digital Hetrodyne Interferometry: A System", Applied Optics, Vol. 19, No. 154 (1980)

[6] Sasaki, O. and Okazaki, H., "Sinusoidal Phase Modulating Interferometry for Surface Profile Measurement", Applied Optics, Vol. 25, No. 18 (1986) pp. 3137-3140

[7] Sasaki, O. and Okazaki, H., "Analysis of Measurement Accuracy in Sinusoidal Phase Modulated Interferometry", Applied Optics, Vol. 25, No. 18 (1986) pp. 3152-3158

[8] Dally, J.W. and Riley, W.F., Experimental Stress Analysis, 2nd ed., McGraw Hill, (1978) p. 455

[9] Stetson, K.A. and Powell, R.L., "Interferometric Hologram Evaluation and Real-Time Vibration Analysis of Diffuse Objects", J. Opt. Soc. Amer., 55 (12), (Dec. 1965), pp. 1694-1695

[10] Abramson, Nils, "Sandwich Hologram Interferometry. A New Dimension in Holographic Comparison", Appl. Opt., 13 (9), (Sept. 1974), pp. 2019-2025

[11] Abramson, Nils, "Sandwich Hologram Interferometry. 3 Contouring", App. Opt., 15 (1), (Jan. 1976), pp. 200-205

[12] Hovanesian, J.D., Hung, Y.Y. and Plotkowski, P.D., "Fringe Compensation in Hologram Interferometry", Experimental Techniques, (Dec. 1981)

[13] Coker, E.G. and Filon, L.N.G., "A Treatise on Photoelasticity" Cambridge University Press (1931)

[14] Frocht, M.M., "Photoelasticity", Vol. 1, John Wiley and Sons, Inc., New York (1941)

[15] Hovanesian, J.D., Qin, Y.W. and Hung, Y.Y., "Multi-wavelength Photoelasticity for Improved Isochromatic/Isopachic Fringe Interpretation", Proc. of Joint SESA/Japan Society for Mechanical Engineers, Hawaii, (May 1982)

17 Liquid crystals applied to photoelasticity

E.J.Hearn & N.Walker

INTRODUCTION

Photoelasticity is a firmly established experimental stress analysis technique whose methodology has remained unchanged for some considerable time. The accurate determination of fringe orders and hence stresses in a model requires many mechanical movements of the basic instrument - the polariscope. Recent developments at Birmingham Polytechnic have shown that the majority of the mechanical movements can be eliminated by introducing electrically controlled liquid crystal cells into the polariscope [1-4]. This type of electrical control is a desirable feature of a prospective automatic polariscope which requires rapid data collection from many points on a photoelastic model. The developments associated with the liquid crystals represent the only major advance in photoelastic instrumentation for over a decade.

The significance of the work has been recognized by British Technology Group who have assisted in filing two patents [5-6]. In each case the advice and practical assistance of the Royal Signals and Radar Establishment (RSRE), Malvern has been recognised by listing personnel from that organisation as co-inventors.

LIQUID CRYSTAL CELLS

Basic cell structure

A simple liquid crystal cell is shown in Figure 17.1. The structure can be simply described as a glass sandwich with a liquid crystal filling. The depth of the liquid crystal is typically 10 micrometers and is achieved by the use of spacers which separate the glass plates. The figure shows mylar strips arranged around the perimeter of the cell. In many commercial applications the glass plates are separated by short (100 μ) pieces of glass fibre dispersed throughout the liquid. These have a negligible effect on the performance of the cell and are considered to be the most effective method of obtaining a uniformity of separation across the full field of view of the cell.

To place an electric field across the cell a thin layer of a transparent conductor (normally tin oxide) is deposited onto the surfaces of the glass. This process is performed by specialist companies who supply batches of coated glass to the cell manufacturers. The manufacturers then deposit a thin layer of a material which can easily be textured. A typical material is polyvinyl alcohol (PVA). When rubbed, the PVA retains a degree of alignment in the direction of the rubbing, so that when the liquid crystal is in contact with the PVA its molecules tend to orientate themselves in the direction of this alignment. On the application of a sufficiently strong electric field across the cell, the molecules lose this alignment and become orientated perpendicularly to the glass plates.

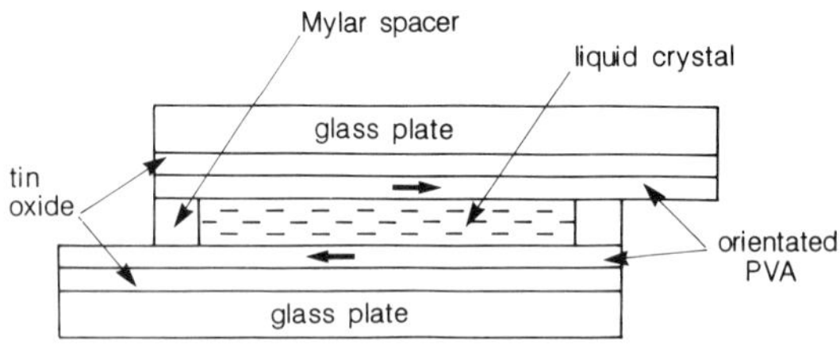

Figure 17.1: Basic construction of liquid crystal cell

Twist cells

Most commercial applications for liquid crystals are in twist cells which are simply cells where the alignment directions on opposite faces are not parallel. The orientation of the liquid crystal molecules then alters gradually from one alignment direction at one face of the cell to another at the other. A twist cell normally is identified by the angle between the alignment directions. Figure 17.2 shows the twisting effect of a 45° twist cell. Polarised light is incident on one surface of the cell so that the plane of polarisation is parallel to the alignment direction. The emergent light is also plane polarised but with a plane of polarisation parallel to the alignment direction in the other surface. Most commercial cells used in displays have a 90° twist and are used as optical switches. The switching process is enacted by applying a sufficiently high voltage so that the molecular units align in a direction perpendicular to the surface of the plates. There is then no anisotropy in the cell and the polarised light is transmitted unaffected.

The 45° twist cells can be used to eliminate the most time consuming mechanical movement in photoelastic studies - the removal and replacing of quarter-wave plates. When positioned between the polarisers and quarter-wave plates as shown in Figure 17.3, the directions of polarisation in the polarisers are effectively rotated to a direction parallel to the fast axes of the quarter-wave plates which consequently have no effect and the polariscope is in the plane-polarised mode. On application of a sufficiently high voltage the twist cells become isotropic and hence ineffective and the quarter-wave plates produce circularly polarised light.

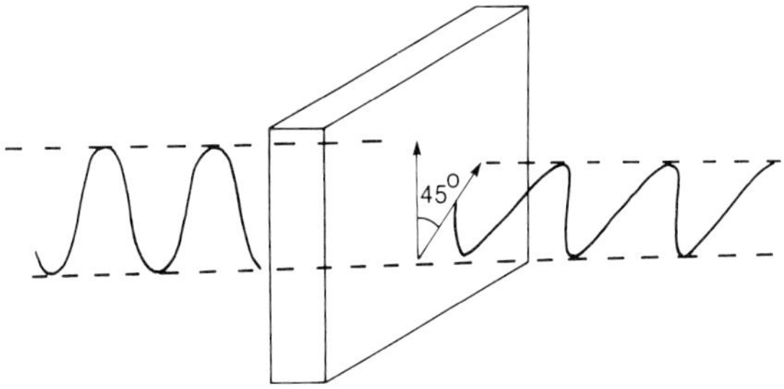

Figure 17.2: Schematic representation of the action of 45° twisted nematic liquid crystal cell

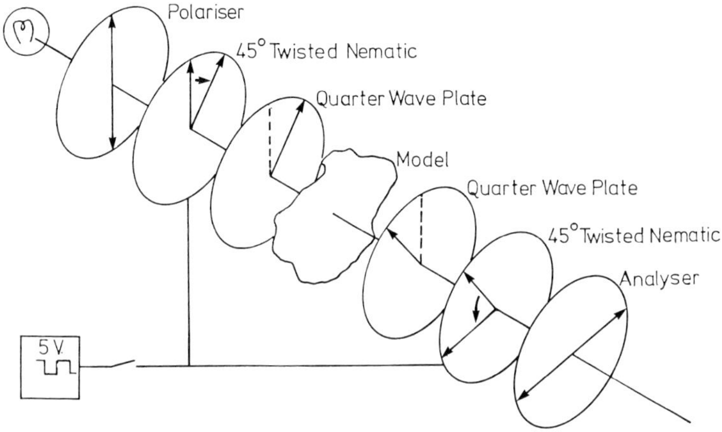

Figure 17.3: Application of twist cells

Freedericksz Cells

In the Freedericksz Cell the alignment directions are set in anti-parallel directions which is the configuration shown in Figure 17.1. This type of cell is seldom used by commercial manufacturers but a novel application as a photoelastic compensator has been identified. Figure 17.4 shows that the retardation in the cell can be tuned by changing the applied voltage. The relationship is non-linear as the molecules are gradually moved from a parallel to perpendicular alignment.

In practice the Freedericksz Cell is positioned adjacent to the photoelastic model and the voltage is altered until the retardation in the cell exactly balances (compensates for) the retardation in the model. The actual value of the retardation in the model can then be assessed from the applied voltage. In this way the integral and partial isochromatic fringe order at a point can be evaluated.

The non-linear relationship between retardation and voltage is easily accommodated by storing the calibration curve in a microcomputer which is also used to control the voltage. Apart from the non-linearity the Freedericksz Cell compensator behaves in the same fashion as the Babinet-Soleil quartz-wedge compensator and is a possible direct replacement for such a device.

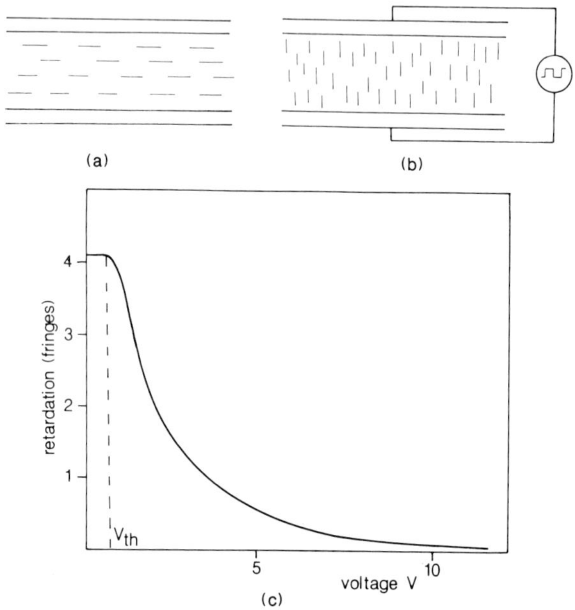

Figure 17.4: Homogeneous Freedericksz Cell

AUTOMATIC POLARISCOPE

Figure 17.5 shows a block diagram of the components of the automatic polariscope under development. The Freedericksz Cell is controlled by a square wave output from a 12 bit digital to analogue converter. The electrical switching of a 10V square wave to the 45° twist cells is easily achieved to give full control of the liquid crystal cells. At present, frozen stress 2D models are used so that a system of three stepper motors can give x y and Q movements of the specimen about a stationary optical axis. Null point in both isoclinic and isochromatic fringe orders is found with a light sensitive detector placed on this axis. Two major problems in development concern optimising the performance of the Freedericksz compensator so that its resolution and repeatability approaches that of the quartz wedge compensators it is intended to replace. In particular the uniformity of the cells must be improved and the effect of temperature quantified.

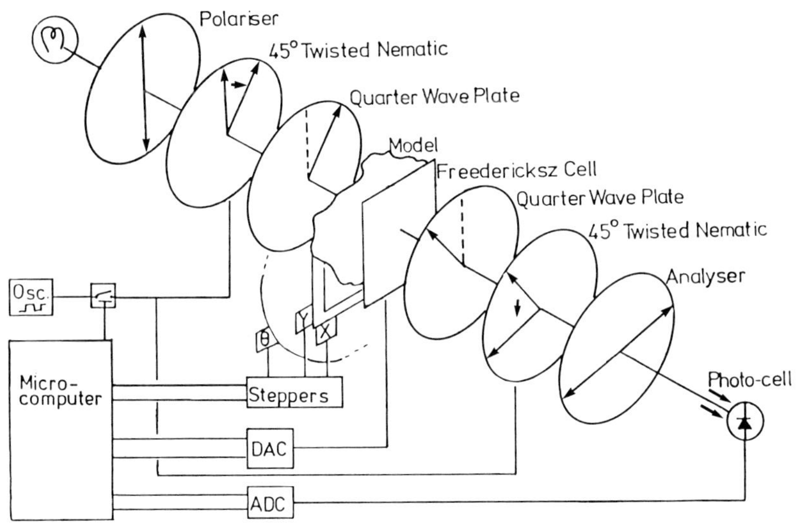

Figure 17.5: Proposed control circuitry

Effect of temperature

The performance of liquid crystal cells is known to be affected by temperature changes. Manufacturers of liquid crystals normally quantify this by describing the movement of the threshold voltage, V_{th} (Figure 17.4). Since most commercial cells are twist cells used primarily as optical switches other considerations are not important. For this reason the performance of the 45° twist cells in the polariscope is not temperature sensitive.

It is important to know whether the Freedericksz Cell calibration curve is affected by temperature. Figure 17.6 shows three specimen calibration curves for a 12μm thick Freedericksz Cell at various temperatures near room temperature. (The liquid crystal becomes isontropic at higher temperature). These curves were produced by placing the cell in a small air oven with PMMA observation windows. The oven was then inserted into a standard polariscope and the calibration curves were obtained with a Babinet-Soleil compensator.

There is an obvious shift to lower retardations as the temperature increases. The sensitivity to temperature is reduced as the voltage is increased and there is little variation in the long 'tail' of the calibration curve. At lower voltages a temperature variation of a few degrees will significantly effect any retardation measurement in the cell and obviously needs to be quantified if the cell is to have accuracy comparable with traditional compensators. Figure 17.7 shows how the voltage required to produce a particular retardation varies with temperature. Alternatives to combine these curves into a single master curve have proved unsuccessful.

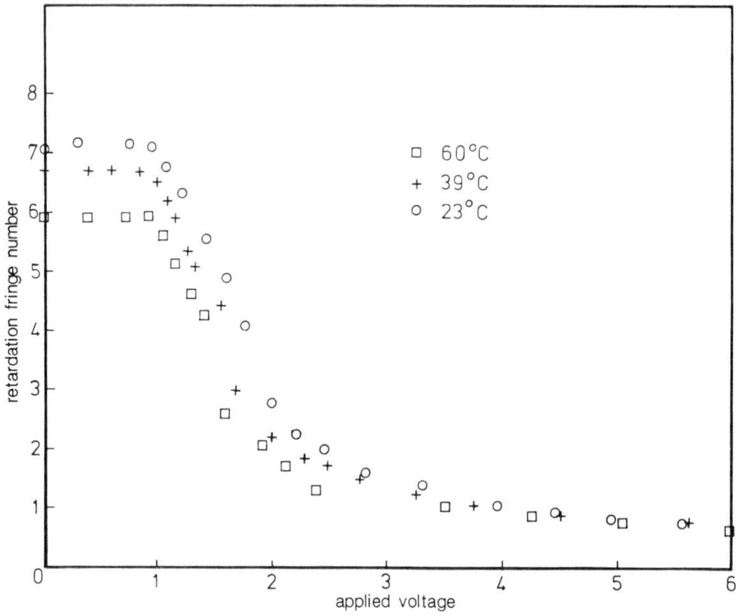

Figure 17.6: *Voltage dependent birefringence of 12μm BDH-E43 Freedericksz Cell*

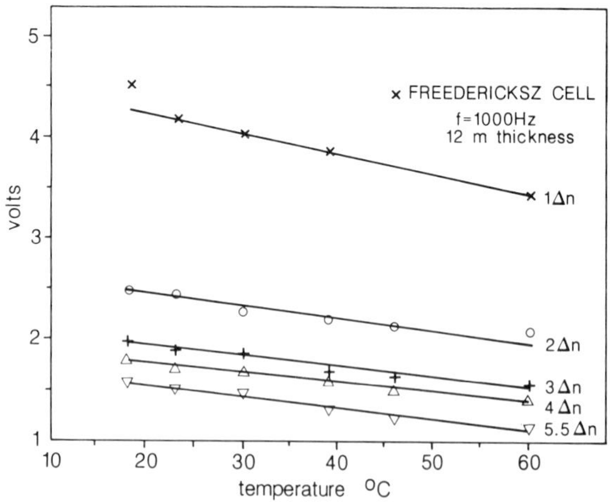

Figure 17.7: *Voltage/temperature dependent birefringence of 12μm BDH-E43 Freedericksz Cell*

Two alternatives exist to account for the temperature effect:

(a) Maintain the cell in a constant temperature environment.

(b) Monitor the temperature of the cell continuously with an unobstrusive thermocouple and load the relevant calibration curve into the computer memory.

The second alternative is under investigation because it is less restrictive on the environment of the cell even the calibration of the cell will be somewhat tedious.

Cell uniformity

Provided twist cells are aligned accurately with the polarizer axes then variations in thickness have negligible effect on the performance of the twist cell. Consequently, commercial cells are made from normal float glass. Unfortunately the retardation in the Freedericksz Cell is linearly dependent on the cell thickness and prototype cells produced in the laboratory are very variable in uniformity of retardation. For the automatic polariscope this is not a significant problem since only the point of the cell lying on the optical axis needs to be calibrated. However, for full field observations the uniformity of the cell needs to be improved. The feasibility of coating optical flats to be used as cell walls is under investigation.

CONCLUSIONS

The development of an automatic polariscope based on liquid crystal cells is proceeding. The switching from plane to circularly polarized light is easily achieved with twist cells and the performance of the Freedericksz Cell compensator is being optimised.

REFERENCES

[1] Hearn, E.J. and Gorman, B., "Some Proposals for the Application of Liquid Crystals to Photoelastic Instrumentation", Strain, 127, 1983

[2] Hearn, E.J., Walker, N. and Shadrokh-Cigari, H., "Further Proposals for the Application of Liquid Crystals to Photoelastic Instrumentation", Strain, 75, 1985

[3] Hearn, E.J. and Walker, N., "Liquid Crystals Applied to Photoelastic Instrumentation", Applied Mechanics Conference, Strathclyde, 1985

[4] Hearn, E.J., Walker, N. and Shadrokh-Cigari, H., "Liquid Crystals Applied to Photoelastic Instrumentation", SEM Spring Conference, Las Vegas, 1985

[5] UK Patent 81 29655, October 1981

[6] UK Patent 84 15129, June 1984

18 Fluorescent photomechanics

M.E.Fourney & B.W.Chang

SUMMARY

In this chapter a new technique for three dimensional stress analysis is demonstrated. It is based on embedding dye molecules in a bifringent material which will fluoresce when properly illuminated. It is found that the increase in the total light output is at least three orders of magnitude greater than that due solely to Rayleigh scattering. Further the light from standard materials used for three dimensional photoelasticity in the past is not due to Rayleigh scattering, but exhibits considerable natural fluorescence. A theory of fluorescent fringe formation is formulated. For randomly oriented dye molecules this predicts a fringe visibility that is less than the ideal for pure Rayleigh scattering, but probably comparable to the actual scattered light photoelastic results. This fringe visibility is significantly improved by the use of a digital image processor.

INTRODUCTION

In the recent past one of the authors has suggested several techniques for three-dimensional stress analysis. These include picosecond holography [1] and scattered-light speckle interferometry [2]. Of these techniques, the scattered-light speckle technique has the greater utility. Even though it has been used to solve problems in both solid mechanics [2,3] and fluid mechanics [4], the scattered-light technique had some shortcomings which limit its usefulness. These are the low light level scattering by interior points in the body and the large amount of light that is scattered by the boundary. Unfortunately the area of interest in mechanics, especially solid mechanics, normally occurs at the boundary. Hence, a weak signal is often sought in a very noisy background.

Fluorescent-photomechanics overcomes these difficulties by casting a fluorescent dye into a model material. The amount of light from a fluorescent model is at least three orders of magnitude greater than that from a normal model exhibiting Rayleigh scattering. Perhaps of more importance is the change of wavelength from the light due to fluorescence which allows boundary scattered light to be filtered out. Experiments using digital image processing on the resulting fringe patterns demonstrate that by using various functions, such as edge enhancement, the quality of the data can be improved significantly.

Two areas have been investigated, first the magnitude and nature of light scattered by various optical model materials was measured. Secondly a qualitative demonstration of the fringe visibility was demonstrated using an image processor. The image processor used is a Gould 8400 with a DEC Vaxstation 200 computer. A monochromator was used to determine the wavelength variation for a constant wavelength input from an argon-ion laser. Three materials were investigated; PMMA, urethene rubber and an epoxy resin (PLM-4). In addition rhodamine 6G dye was embedded in the epoxy resin and urethene rubber for evaluation of the fluorescence effect.

The PMMA shows mostly Rayleigh-type scattering, with very low intensity, no shift in wavelength and very little broadening. Both the urethene rubber and the epoxy resin scattered much more light. However both also have a shift in peak output wavelength and significant broadening. It is important to note that these materials have been extensively used for scattered-light photoelasticity in the past. The light scattered due to fluorescence in epoxy resin has a peak intensity of 276 times over that of PMMA. Further, the total light received (area under the curve) for the fluorescent material is several thousand times greater than from PMMA.

DESCRIPTION OF THE FLUORESCENT PROCESS

The fluorescent process can be explained with the aid of a schematic energy level diagram of a fluorescent dye molecule (Figure 18.1). When a fluorescent molecule is illuminated with incident light which has a wavelength within the absorption band, an electronic transition will take place from the ground state S_o to the higher energy singlet states S_n. For singlet states other than S_1, the excess energy is converted to heat by the vibrational decay from S_n to S_1. Subsequent transitions from the singlet state S_1 to the ground state S_o result in the emission of radiation known at fluorescence. The emitted radiation is red shifted, i.e. the re-emitted light has a longer wavelength than the incident light. At this stage the emission is spontaneous, and therefore is incoherent.

There are several factors which determine the polarization of the emitted fluorescent light namely: (1) the orientation of the absorption and fluorescent dipole moment of the fluorescent molecule, (2) secondary fluorescence, (3) rotational Brownian motion of the excited dye molecule during the transitional lifetime and (4) the physical orientation of a collection of dye molecules may be random.

In general the absorption dipole moment axis of the fluorescent molecule is not parallel to the fluorescent dipole moment axes but dependent upon the wavelength of the incident ratiation

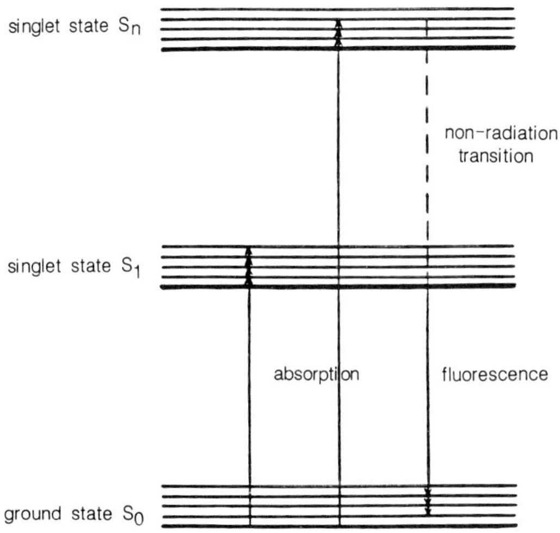

Figure 18.1: Energy level diagram of fluorescent dye molecules

and the dye molecule structure. However, for large ring-type fluorescent molecules such as Rhodamine 6G (which was used in this study), the transition moment of the main absorption band is oriented parallel to the long axis of the molecule and by proper selection of the illuminating wavelength, the absorption and fluorescent dipole moment axis are parallel. Likewise, by proper control of the dye molecule concentration (below 0.01 grams/litre) the secondary fluorescence effect can be minimized.

Although there is a significant transition lifetime (approximately 1 to 10 nanoseconds) between the absorption and re-emission process, when the molecule is cast into the polymer model the rotational relaxation time is much longer and the dye molecule may be considered as immobile. Finally, the dye molecules may be introduced into the polymer system by homogeneous dispersion, or by a direct chemical incorporation into the polymer chain. In this study the dispersion technique was used resulting in a random orientation of the dye molecules.

MODEL PREPARATION

Several birefringent materials were used as host for the dye molecules. This study reports on (i) gelatin, (ii) PLM-4 epoxy resin from Photoelastic Inc., and (iii) Solithane 113 urethane rubber from Thiokol Chemical Corporation. In all cases the dye used was Rhodamine 6G produced by Kodak.

Gelatin was chosen because of its ease of casting, low cost and high birefringent sensitivity. A mixture of 100 grams of gelatin powder in 1600ml of glycerin with a dye concentration of 0.01 grams/litre was used. Both PLM-4 and Solithane were cast following the manufacturers recommendations with a dye concentration of 0.01 grams/litre. The PLM-4 required a curing cycle with temperatures that were high enough to cause the dye molecule to become bleached. This happens becaue the dye molecules reach a highly reactive state due to the thermal excitation and react chemically with the solvent molecules, dissolved oxygen or impurities to yield decomposed products. Although there are methods of avoiding this problem the solutions were sufficiently difficult to discourage further use of this material.

Solithane 113 is a urethane rubber with a relatively low curing temperature and it is very birefringent. It was successfully cast with a dye concentration of 0.01 grams/litre.

SPECTROSCOPIC STUDIES

It was noted that with all of these materials a degree of natural fluorescence occurs without the introduction of the dye molecule. Therefore a study was undertaken to determine the spectral character of the re-emitted light for these materials.

The re-emission spectrum of the various materials was determined using the experimental setup shown in Figure 18.2. The model material was illuminated with a beam of light from an argon laser, a single line of wavelength 514nm. A good quality monochromator with a slit width of 3 angstroms was used in all cases except one. This narrow band of wavelength greatly reduced the intensity of the re-emitted light. This intensity was measured by RCA C 31034 photomultiplier tube. A discriminator eliminates the dark pulses, a timer allows photon counting for a fixed interval and a rate meter formed the readout.

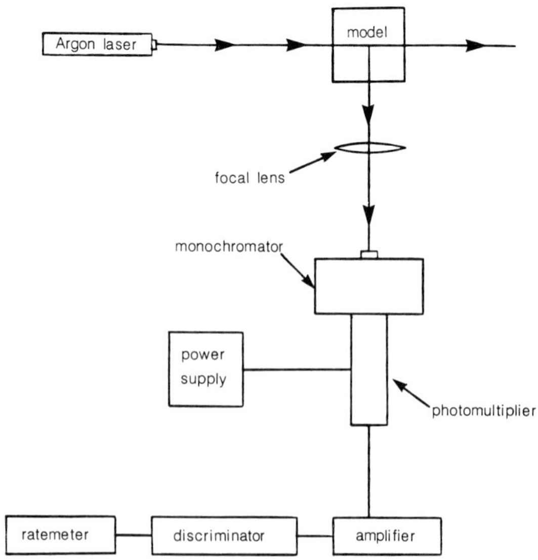

Figure 18.2: Experimental arrangement for studying model re-emission

The results of these tests are shown in Figures 18.3 - 18.5. Examination of the re-emission spectrum of neat PLM-4 and Solithane 113 shows that for both of these materials the curves are red-shifted and significantly broadened. By contrast PMMA (Polymethylmethacrylate) (Figure 18.5) exhibits mostly Rayleigh scattering with no significant red-shifting or broadening. However the intensity of the scattered light is very low. Note that although the peak count rate on Figures 18.3 and 18.5 is comparable, the input power for the PMMA is 50 times higher and a monochromater slit width of 5 angstroms was required to yield good quality measurements.

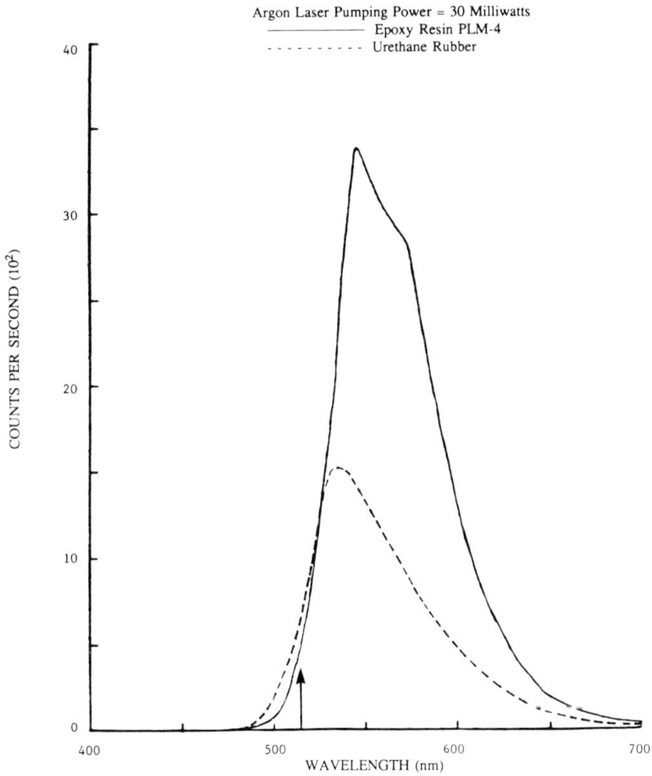

Figure 18.3: *Re-emission spectrum of epoxy resin PLM-4 and urethane rubber*

When dye molecules were added to epoxy resin (in this case the epoxy resin was not fully cured to minimize the bleaching problem) the resulting light intensity increased by two orders of magnitude. Therefore the gain in light intensity for the fluorescent material over PMMA, which has been used by one of the authors [2] in scattered light speckle interferometry studies, is four orders of magnitude. This tremendous gain came at a

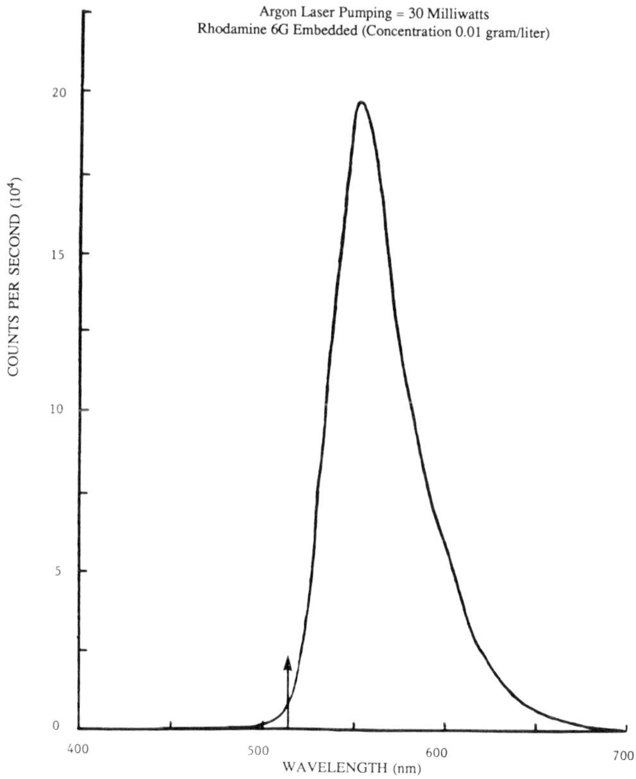

Figure 18.4: *Re-emission spectrum of epoxy resin PLM-4 with fluorescent dye embedded*

cost of a much broader re-emission spectrum and in the present case at a loss of optical coherence. Similar tests on the PLM-4 with dye molecules have not yet been completed. However, qualitative comparison of model results discussed later in this chapter indicate that the re-emission spectrum is similar to that of the epoxy resin.

THEORY OF FRINGE FORMATION

The relationship of the fluorescent light intensity of a birefringent fringe to the stress state in the medium in the most general case is quite complex. Only the case in which the secondary principal stress directions are constant along the illuminating light path will be considered here.

Consider a fluorescent dye molecule in a stressed medium with absorption and fluorescent dipole moments coincide whose unique symmetry axis (long axis of dye molecule) denoted by D which can be located in a laboratory based coordinate frame

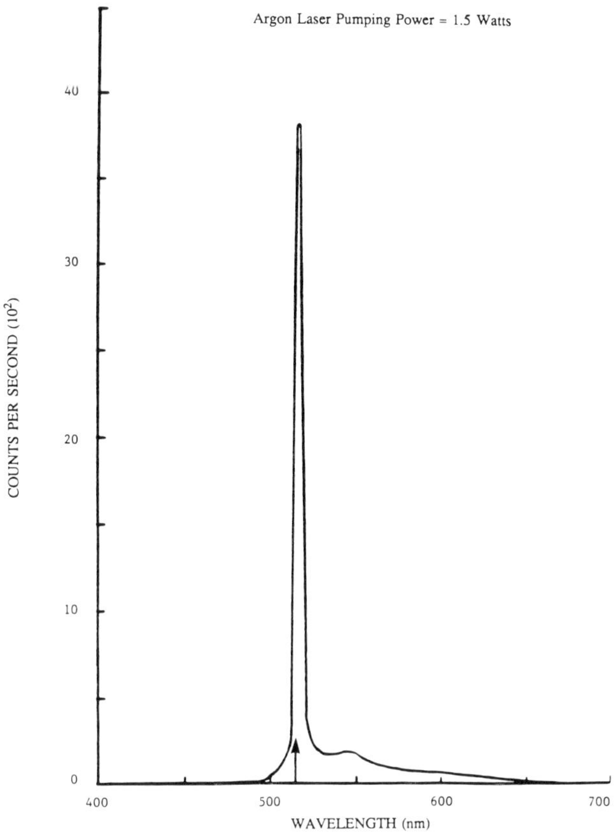

Figure 18.5: Re-emission spectrum of PMMA

O-$X_1 X_2 X_3$ by polar and azimuthal angle θ, ϕ as in Figure 18.6. When a wave of incident light passes through a dyed transparent birefringent material along the X_2 axis, this incident light will change its polarization characteristic in the plane of the wave front, as result of stress state. Generally it can be described as elliptically polarized (Figure 18.7), and the equation of this ellipse is

$$\frac{U^2}{a^2} + \frac{V^2}{b^2} - \frac{2UV}{ab} \cos \Delta = \sin^2 \Delta \qquad (18.1)$$

where $U = a\cos\omega t$ and $V = b\cos(\omega t + \Delta)$ are two components of electric vector oscillation along the directions of secondary principal stresses σ_p, σ_q in the transverse plane and Δ is the relative phase difference between these two oscillations.

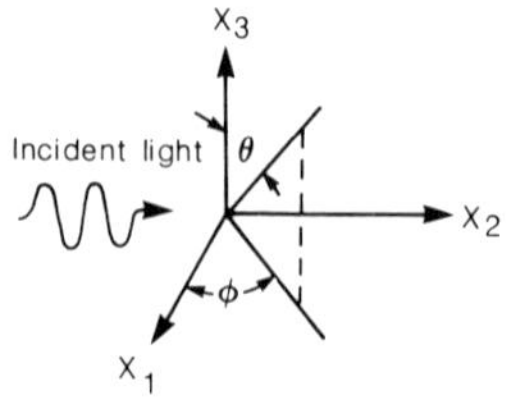

Figure 18.6: Orientation of a fluorescent dye molecule with a unique axis D in the laboratory frame $O\text{-}X_1X_2X_3$

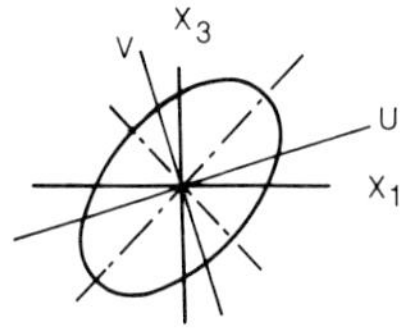

Figure 18.7: Light ellipse and direction of the secondary principal stresses

Since the dye molecule will only absorb radiation along a certain direction D, it is necessary to calculate the radiation component of incident light along the direction D by first determining the incident light radiation components along the laboratory coordinate frame. It can be shown as

$$\begin{matrix} E_1 \\ E_2 \\ E_3 \end{matrix} = \begin{matrix} \cos\alpha & 0 & -\sin\alpha \\ 0 & 1 & 0 \\ \sin\alpha & 0 & \cos\alpha \end{matrix} \quad \begin{matrix} U \\ 0 \\ V \end{matrix} \tag{18.2}$$

where α is the angle between the direction of secondary principal stress σ_p and laboratory coordinate axis X_1 (Figure 18.8). Next, project these laboratory coordinate components onto the direction of the absorption dipole moment of the fluorescent dye molecule thus,

$$E_D = [\ \sin\theta\cos\phi \quad \sin\theta\sin\phi \quad \cos\theta\ ] \quad \begin{matrix} E_1 \\ E_2 \\ E_3 \end{matrix} \tag{18.3}$$

which can be written as

$$E_D = R\cos(\omega t + \psi_1) \tag{18.4}$$

where R and ψ_1 are the resultant amplitude and the phase angle respectively.

$$R = (a^2\cos^2\alpha + b^2\sin^2\alpha)\sin^2\theta\cos^2\phi + 2(a^2 - b^2)\cos\alpha\sin\alpha\sin\theta\cos\theta\cos\phi + (a^2\sin^2\alpha + b^2\cos^2\alpha)\cos^2\theta + 2\,ab\cos\Delta\,[(\cos^2\alpha - \sin^2\alpha)\cos\theta\sin\theta\cos\phi + \cos\alpha\sin\alpha(\cos^2\theta - \sin^2\theta\cos^2\phi)]^{\frac{1}{2}} \tag{18.5}$$

and

$$\psi_1 = \frac{\text{arc tan } b \sin\Delta(\cos\theta \cos\alpha - \sin\theta \cos\phi \sin\alpha)}{a(\sin\theta \cos\phi \cos\alpha + \cos\theta \sin\alpha) + b \cos\Delta(\cos\theta \cos\alpha - \sin\theta \cos\phi \sin\alpha)} \tag{18.6}$$

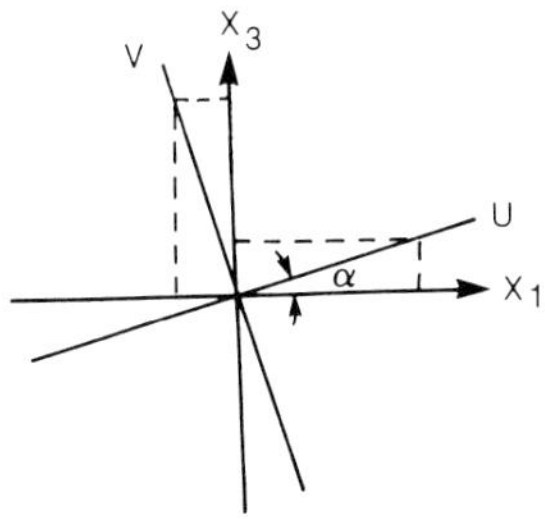

Figure 18.8: *The incident light radiation components along the laboratory coordinate frames for ease without rolation of the secondary principal axes*

For an immobile fluorescent dye molecule, whose absorption and fluorescent dipole moments are parallel, the resultant fluorescent light intensity $I^F_{\theta\phi}$ at a fluorescent molecule is proportional to the square of the resultant amplitude of absorption, and the resultant fluorescent light will be polarized along the direction D as well. Then

$$I^F_{\theta\phi} = K_1 R^2 \tag{18.7}$$

where K_1 is a constant depending on the quantum yield of the fluorescent molecule. The intensity of the fluorescent light components which are polarized along each laboratory coordinate axes can be determined as

$$\begin{Bmatrix} I^F_1 \\ I^F_2 \\ I^F_3 \end{Bmatrix} = I^F_{\theta\phi} \begin{Bmatrix} \sin^2\theta\cos^2\theta \\ \sin^2\theta\sin^2\phi \\ \cos^2\theta \end{Bmatrix}_{\theta\phi}$$

where I^F_1, I^F_2, and I^F_3 correspond to the axes X_1, X_2, and X_3 respectively.

Because this fluorescent radiation is incoherent, the overall fluorescent light intensity is the total intensity of all excited dye molecules. For a randomly oriented collection of fluorescent dye molecules, the probability of every polar and azimuthal angle

θ and ϕ is equal. These two angles have all possible values from 0 to π and 0 to 2π respectively, thus, the function of angle θ and ϕ can be replaced by their average values. Then the fluorescent light intensity is given by

$$\begin{bmatrix} I_1^F \\ I_1^F \\ I_3^F \end{bmatrix} = \frac{1}{2\pi} K_2 \int_0^{2\pi} \int_0^{\pi} \begin{bmatrix} I_1^F \\ I_2^F \\ I_3^F \end{bmatrix}_{\theta\phi} \sin\theta \, d\theta \, d\phi \qquad (18.9)$$

where K_2 is a combined constant of instrument, concentration and volume factors.

When the incident light passes through dyed model materials, there is absorption. For a place wave, the incident light intensity along the light path will not be constant but will decrease as $e^{-\beta X_2}$, (where β is the absorption coefficient). This is the expotential law of absorption. The equations of the fluorescent light intensity may be modified to take account of absorption by multiplying the intensity by the factor $e^{-\beta X_2}$. For complete expressions

$$I_1^F = K_3[(a^2+b^2) + 2(a^2 \cos^2\alpha+b^2 \sin^2\alpha)-4ab \cos\alpha \sin\alpha \cos\Delta]e^{-\beta x_2} \qquad (18.10)$$

$$I_2^F = K_3(a^2 + b^2)e^{-\beta x_2} \qquad (18.11)$$

$$I_3^F = K_3[(a^2+b^2) + 2(a^2\sin^2\alpha+b^2 \cos^2\alpha) + 4ab \cos\alpha \sin\alpha \cos\Delta]e^{-\beta x_2} \qquad (18.12)$$

where $K_3 = \frac{2}{15} K_1 K_2$

For the fluorescence experiment performed in the right angle geometry, it is convenient to assume the direction of observation is along the laboratory X_3 axis. Two components of fluorescent light intensity I_1^F and I_2^F will be observed. The intensity I_2^F, which contains no information regarding the relative phase retardation, could be eliminated by using an analyzer. The total observed fluorescent light intensity I_F can be expressed by

$$I_F = I_1^F + I_2^F = K[(a^2 + b^2) + (a^2 \sin^2\alpha + b^2 \cos^2\alpha)$$
$$- 2ab \sin\alpha \cos\alpha \cos\Delta]e^{-\beta x_2} \qquad (18.13)$$

where constant K is equal to $2K_3$.

STRESS OPTIC-LAW

The relationship between the secondary principal stresses and the optical relative retardation which has been used in standard scattered-light photoelastic techniques is assumed to apply to the fluorescent light techniques. The equation governing this relationship is given by

$$\sigma_p - \sigma_q = \frac{1}{C}\frac{d\Delta}{dX_2} \tag{18.14}$$

where C is the stress-optical coefficient, dX_2 is an increment of distance along the optical path, and $d\Delta$ is the relative retardation due to stress in moving from point X_2 to $X_2 + dX_2$. The standard scattered-light photoelastic intensity may be expressed as:

$$I_s = K^* (a^2 \cos^2\alpha + b^2 \sin^2\alpha - 2ab \sin\alpha \cos\alpha \cos\Delta) \tag{18.15}$$

When this is compared to the light intensity due to fluorescence (equation (18.13)) it will be seen that they are quite similar. The fluorescent light has a stronger background intensity and, as will be seen, decreases the fringe visibility . In order to compare the two equations it will be helpful to consider the case when the direction of observation and the axes of the light ellipse both bisect the angle between the two secondary principal stress directions. This is the situation that yields the sharpest contrast to the fringe patterns. The light intensity expression of equations (18.13) and (18.15) reduce to:

$$I_F = K a^2(3 - \cos\Delta) \tag{18.16}$$

$$I_s = K^* a^2(1 - \cos\Delta) \tag{18.17}$$

It will be noted that the expotential decay factor has been omitted. The relative intensity of the fluorescence varies between 2 and 4 and the fringe visibility, defined as:

$$V = \frac{I_{max} - I_{min}}{I_{max} + I_{min}} \tag{18.18}$$

has a value of 1/3. For the scattered-light technique the intensity varies between 0 and 1, and the fringe visibility is 1. This would seem to favour the scattered-light technique. However two points are worth noting. First, the equation for the scattered-light intensity assumes that light which is scattered perpendicular tc the incident beam is plane-polarized and has the same wavelength as the incident light. This type of scattering is known as Rayleigh scattering. However, for the standard materials used for scattered-light photoelasticity, this occurs only when the incident light is not within the absorption band of the material. Otherwise as noted by Gross-Petersen [5] the re-emitted light is a combination of Rayleigh scattering and a component which he describes as a DC-term, but we describe as the material's natural fluorescence. In any event the fringe visibility is considerably less than unity. Secondly the tremendous increase in intensity in the fluorescent technique allows one to use various filtering techniques to improve the fringe visibility.

EXPERIMENTAL INVESTIGATION

A demonstration was conducted to obtain a stress-induced fringe pattern. The model was a 4.5 inch cube made of Solithane 113. The material was made of 7pbw Solithane 113 and 5pbw curing agent with 0.001 grams/litre Rhodamine 6G and cured as per the manufacturer's instructions. The model was loaded by a line load to yield the Boussinesq problem. A schematic of the experimental arrangement is shown in Figure 18.9. As indicated, illumination is by means of a linearly polarized beam of light from an argon-ion laser (λ = 514nm).

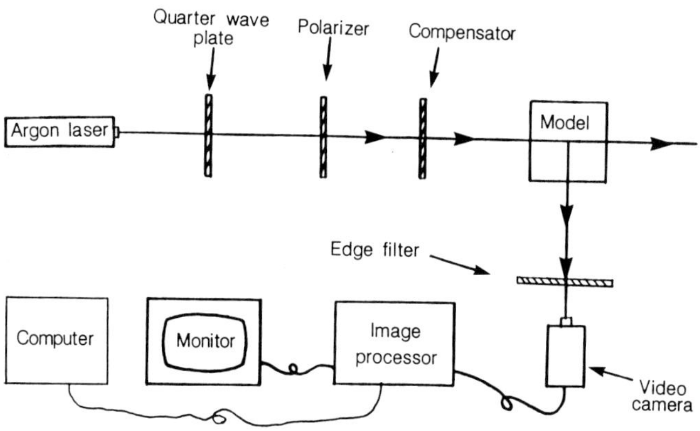

Figure 18.9: Fluorescent light polariscope and digital image processing system

The fluorescent light birefringent fringe pattern and the typical intensity variation along the light path is shown in Figure 18.10. The fringe visibility is quite low and in order to overcome this difficulty a digital image processor (a Gould IP-8400, with a DEC VAXstationII computer) is employed. The image is processed in two ways. First the absorption exponential decay is removed. Secondly, the image visibility is significantly enhanced by various filtering procedures. The results of two of these are shown in Figure 18.11. The upper portion of this Figure shows the image of the fringe pattern in which contrast has been linearly stretched. The lower portion shows the results when a histogram equalization scheme is used.

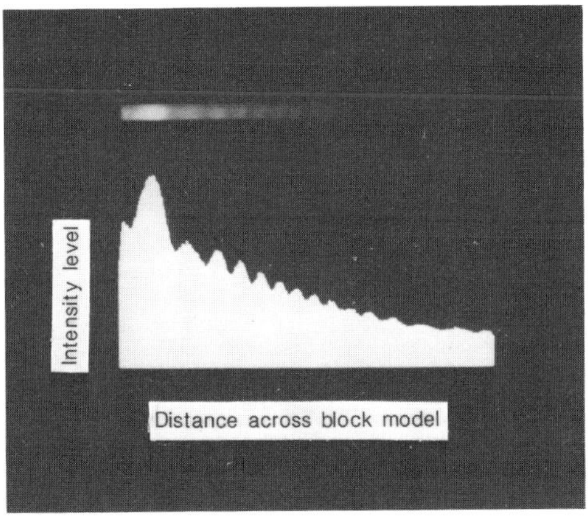

Figure 18.10: *Photograph of fluorescent light fringes image and typical intensity variation along the light path*

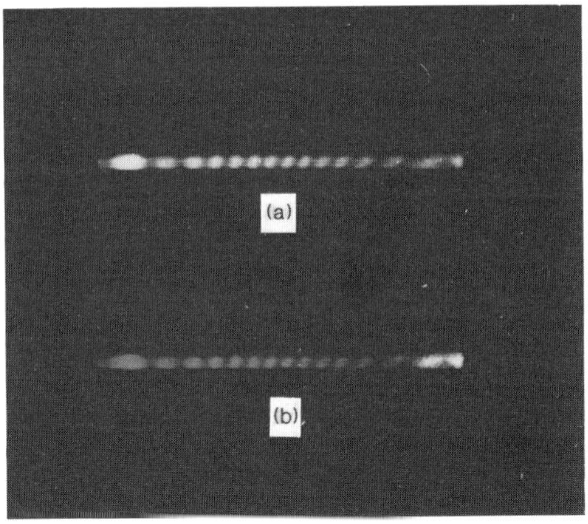

Figure 18.11: *Photograph of enhanced fluorescent light fringes results of contrast stretch techniques (a) linear contrast stretch (b) histogram equalization*

DISCUSSION

We have shown that it is possible to embed fluorescent dye molecules into photoelastic materials. The increase in total amount of available light over materials which exhibit Rayleigh scattering is considerable. Also the shift in wavelength due to the fluorescence process is utilized to filter out unwanted scattering. The broadened re-emission spectrum and loss of optical coherence are the negative aspects of this technique.

The mathematics of a fringe formation theory are presented and some experiments have been performed. The fringe visibility in these tests was not remarkable. However with the aid of a digital image processor, fringes of good definition were obtained. The methods need further investigation before any further definitive statements can be made.

REFERENCES

[1] Fourney, M.E. and Barker, D.B., "Picosecond Holography", Applied Physics Letters, Vol. 21, No. 1, July 1, 1972

[2] Barker, D.B. and Fourney, M.E., "Using Scattered-Light Speckle Interferometry", Experimental Mechanics, Vol. 16, No. 5, May 1976

[3] Barker, D.B. and Fourney, M.E., "Three-Dimensional Interferometric Investigation of Stress-Intensity Factor along a Crack Front", Experimental Mechanics, Vol. 17, No. 7, July 1977

[4] Barker, D.B. and Fourney, M.E., "Measuring Fluid Velocities with Speckle Patterns", Optical Letters, Vol. 1, No. 4, Oct. 1977

[5] Gross-Petersen, Jorgen, "A Compensation Method in Scattered Light Photoelasticity", The Photoelastic Effect and Its Application, IUTAM Symposium Ottignies/Belgium, Sept. 1973

19 The application of holography to the deformation of a piston

W.T.Evans & R.M.Russell

SUMMARY

The thermal distortion of a piston is an important factor which influences its performance and the main method used to date for estimating this distortion consists of finite element modelling.

The current study is concerned with measuring the thermal distortion experimentally using holographic interferometry which is well suited for the purpose, being a non-contacting field method. In order to find the absolute displacement at a given point on the piston, three displacement vectors must be constructed at that point and to this end three separate holograms were simultaneously obtained from different viewing positions. Photographs were taken through each of the resulting holograms, the fringe patterns digitised and used as input to a computer program which produced the final displacement patterns.

INTRODUCTION

The overall objective of the study was to measure the thermal distortion of diesel engine pistons using holographic interferometry and compare the results with those predicted by the finite element method. The hot shape of a piston is a critical factor in its performance and a great deal of effort has been expended by designers to control this shape [1]. Bimetallic pistons such as those being studied incorporate a steel band which is cast into the piston during manufacture. The differential expansion of the aluminium and steel produces a locked in tensile stress in the aluminium piston skirt. As the piston is heated in operation this stress must first be relaxed before the skirt is able to expand freely, and control is thus established over the hot shape of the skirt. The geometry of the steel insert, among other design features, affects the amount of control it can exert over the skirt.

Holography is ideal for measuring the expansion of piston skirts over their whole surface and thus provides a means by which different methods of expansion control can be compared.

Several holographic methods have been developed which include

Fringe localisation (FL) [2], Fringe counting (FC) [3] and Zero Fringe (ZF) [4]. The most accurate of these methods was considered to be the ZF method [5] which had the disadvantage of requiring three holograms compared with one hologram using the other methods. An experimental comparison between the techniques based on a single hologram analysis [6] showed that the FC method gave a maximum percentage error of 9% whereas an adaption of this method [7] gave a maximum percentage error of 2%. Adaptions of the ZF method have been showed by the authors [8] to have a maximum percentage error of 5% and other workers [9] have reported a 2% accuracy.

Wellworthy Ltd. supplied pistons to be measured and also finite element solutions for the thermal expansion due to the temperature distribution obtained from the holographic analysis.

BASIC THEORY

The phase changes resulting from surface deformation can be determined by examination of the fringe patterns and related to the vector displacement of surface points by applying principles of classical interferometry. The method of analysis used the classical principles of interferometry first proposed by Ennos [4].

Figure 19.1 shows a point P on the object surface illuminated by a laser beam whose direction is specified by the illumination unit vector I . A hologram is taken of the surface before deformation, the hologram processed and precisely repositioned in its holder. The object surface is then displaced so that point P moves to position P^1, a small distance D from P in some arbitrary direction. On looking through the hologram H1 the observer will see the reconstructed wavefront from P superimposed on the 'live' interference holographic fringes. If the direction of viewing through the hologram is specified by the unit vector $\underline{O}_1$, and the unknown vector displacement $\underline{D}$ produces a difference in fringe order of N_1 at the point P then the relationship between these variables is given by

$$(\underline{O}_1 - \underline{I}) . \underline{D} = N . \lambda \tag{19.1}$$

where λ = wavelength of the laser light Equation 19.1 shows that the holographic interference fringes will indicate only that component of the vector displacement $\underline{D}$ which lies in the direction of the sensitivity vector $(\underline{O}_1 - \underline{I})$. a direction which bisects the angle between the illumination and observation directions. In order to find the cartesian components of displacement Dx, Dy and Dz, three viewing positions must be made producing three independent vector components of $\underline{D}$, each having its corresponding fringes order values N_1, N_2 and N_3 respectively. The additional equations required will be:

$$(\underline{O}_2 - \underline{I}) . \underline{D} = N_2 \lambda \tag{19.2}$$

$$(\underline{O}_3 - \underline{I}) . \underline{D} = N_3 . \lambda \qquad (19.3)$$

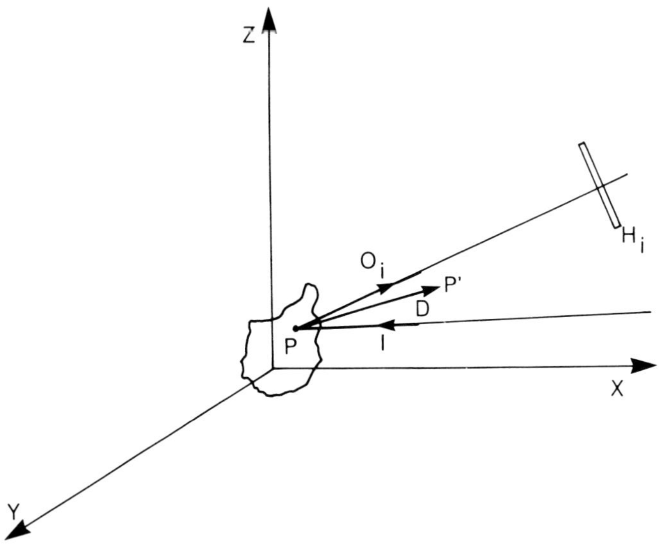

Figure 19.1: Path of incident and reflected rays at point P

On substituting the direction cosines for each of the unit vectors $\underline{O}_1$, $\underline{O}_2$, $\underline{O}_3$ and $\underline{I}$:

$$\underline{O}_1 = \cos\alpha_1 + \cos\beta_1 + \cos\gamma_1$$
$$\underline{O}_2 = \cos\alpha_2 + \cos\beta_2 + \cos\gamma_2$$
$$\underline{O}_3 = \cos\alpha_3 + \cos\beta_3 + \cos\gamma_3$$

Rewriting Equations 19.1 - 19.3 in matrix form:

$$\begin{matrix} (\cos\alpha_1-\cos\alpha_I) & (\cos\beta_1-\cos\beta_I) & (\cos\gamma_1-\cos\gamma_I) \\ (\cos\alpha_2-\cos\alpha_I) & (\cos\beta_2-\cos\beta_I) & (\cos\gamma_2-\cos\gamma_I) \\ (\cos\alpha_3-\cos\alpha_I) & (\cos\beta_3-\cos\beta_I) & (\cos\gamma_3-\cos\gamma_I) \end{matrix} \quad \begin{matrix} D_x \\ D_y \\ D_z \end{matrix} = \begin{matrix} N_1 \\ N_2 \\ N_3 \end{matrix} \lambda \qquad (19.4)$$

When the coordinates of the illuminating and viewing positions are known then the geometric matrix can be obtained. The viewing sequence will provide values of N_1, N_2 and N_3 at each point and hence Equation 19.4 can be solved to provide values of Dx, Dy and Dz at each point.

EXPERIMENTAL DETAILS

Two large iron tables formed the optical bench, the upper table resting on air springs mounted on the lower table and a 50mW He-Ne laser used as the light source. The optical set up is shown schematically in Figure 19.2 where the original beam is subdivided to provide object and reference beams for each of the three holograms. Two of the holograms were located on either side of the piston and at the same vertical height, the third hologram being raised above the other two. A mirror was inserted either side of the piston in order to cover the maximum piston scan area. Thus each of the holograms on either side of the piston could view the piston directly, together with a mirror image. Plate holders were made which provided three edge point contact and three face point contact between the holographic plate and holder. The piston was heated using an electrical heater embedded in the crown of the piston and temperatures were monitored using twelve thermocouples linked to a computer via a suitable interface which allowed temperature sweeps to be taken automatically every three seconds. Three 16mm movie cameras were used to record the live fringes via an electronic trigger device which allowed a simultaneous start for all three cameras together with a time interval selection unit for selecting the required number of exposures per second.

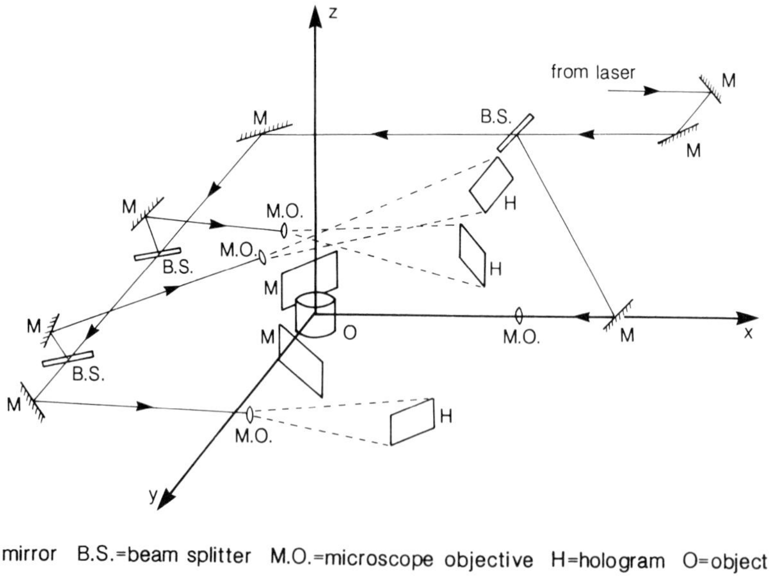

Figure 19.2: Schematic layout of the hologram bench

In order to obtain results from as wide as temperature range as possible, the initial holograms were exposed at a temperature mid-way between the initial and final temperatures of the actual test. After processing, the holograms were positioned in their respective holders and the temperature stabilised at the initial lower temperatures of the heating sequence. The films then recorded the live fringes as the temperature was gradually raised to its final value. At certain intervals during the heating sequence, marker cards were inserted into the viewing area so as to be visible from each camera position, and a temperature sweep was recorded each time a marker card was used. The resulting fringes on the film thus showed a high fringe density at the start of the heating sequence which decreased to zero fringe density at the initial hologram exposure temperature. As the temperature further rose to the final temperature of the test the fringe density again increased.

COMPUTER PROGRAM

The developed films clearly showed the line fringe formation from each viewing station together with the time marker cards. Photographs were printed from each viewing station showing fringes which had been recorded simultaneously and these were analysed using a computer program written to obtain the displacement vectors from the digitised fringe centres [8]. In addition to the fringe photographs the program requires the global coordinates of the front nodal point of each camera lens, the laser spatial filter position and the piston diameter and position.

The program contains a set of subroutines, Figure 19.3, each of which completes a particular part of the analysis and which can be viewed in either numerical or graphical form [10]. It transforms the digitised fringe positions to three dimensional global coordinates which are transferred back to the piston surface via the camera nodal lens position. The geometric matrix in Equation 19.4 is then evaluated to provide the final displacement vectors which can be viewed from any spatial position. Figure 19.3 shows the flow chart for the program.

FINITE ELEMENT COMPARISON

Wellworthy Ltd. supplied the finite element results based on the temperatures measured during the experimental run corresponding to the movie camera recording. Their first task was to apply thermal boundary conditions to the piston surfaces to reproduce the temperature gradient throughout the piston. When this had been achieved, the deformations were calculated.

Figure 19.4 indicates the position of the thermocoupler on the piston surface together with the initial experimental temperatures and final experimental/predicted temperature. A general sketch of the finite element mesh used by Wellworthy Ltd. is

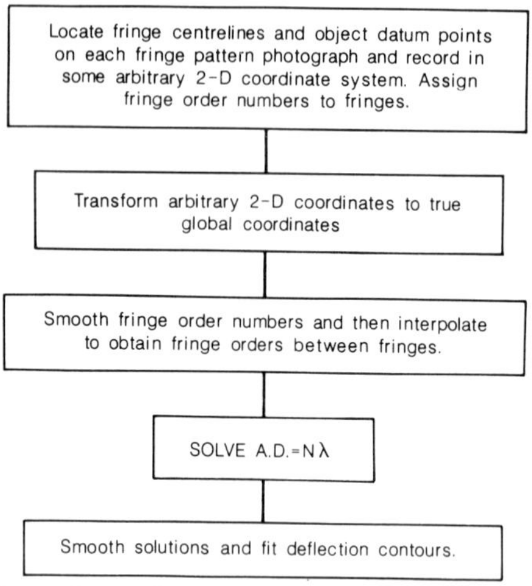

Figure 19.3: Flow chart for computer program

shown in Figure 19.5.

RESULTS

Figure 19.6 shows a plot of the axial movement of the piston along a line on the outside of the piston at an angular position $\theta = 50°$ (Figure 19.4). The piston was bolted on to a vertical column which extended to the underside of the crown at a position z = 25mm relative to the centre line of the cross hole and the results are plotted relative to this position. The holographic plot shows a more linear distribution than that predicted by finite elements and work is currently centred on comparing radial movements for the temperatures shown in Figure 19.4.

ACKNOWLEDGEMENT

This work was part sponsored by an SERC grant. The finite element results and pistons were supplied by Wellworthy Ltd., whose co-operation is gratefully acknowledged.

REFERENCES

[1] Robinson, J.E., "Engine Design Experiences and Development of Critical Components", Proc. I. Mech. E., Automobile Div., 193 (28), 1979, pp. 207]26

[2] Haines, K.A. and Hildebrand, B.P., Appl. Opt., 5, 1966, pp. 172-3

[3] Aleksandrov, E.B. and Bonch-Bruevich, A.M., Zh. Tekh. Phys., 37, 1967, pp. 360-60 (Translation in Sov. Phys. Tech. Phys., 12 (1967), pp. 258-65

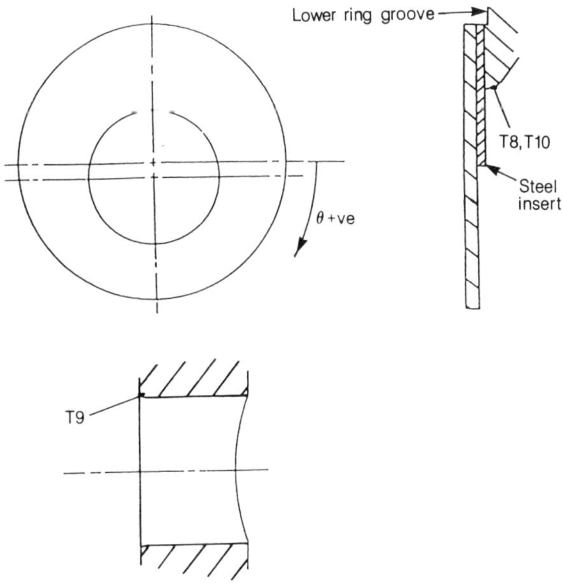

Part number XLMN090220STND						
Thermo-couple	Theta (deg)	Z (mm)	Radius (mm)	Initial Temp(°C)	Final Temp(°C)	Predicted Temp(°C)
T1	45	bottom of bowl		28.2	50.7	44
T2	0	69.7	o/r	28.2	40.1	43
T3	90	41.1	o/r	28.2	40.4	42
T4	0	27.2	o/r	28.2	39.8	42
T5	90	27.2	o/r	28.2	38.3	41
T6	0	-50.8	o/r	28.2	37.4	36
T7	90	-50.8	o/r	28.2	37.3	36
T8	90	see above		28.2	39.7	38
T9	0	see above		28.2	39.8	41
T10	270	see above		28.2	40.0	38
T11		centre of u'crn.		28.2	41.2	44
T12	50	0	o/r	28.2	38.7	39

Figure 19.4: Measured and predicted temperatures

[4] Ennos, E.A., "Measurement of In-plane Surface Strain by Holographic Interferometry", J. Phys. E., 1, 1968, pp. 731-34

[5] Briers, J.D., "The Interpretation of Holographic Interferograms", Opt. 2 Quantum Electronics, 8, 1976, pp. 469-501

[6] Pryputniewicz, R.J. and Bowley, W.W., "Techniques of holographic displacement measurement, an experimental comparison", App. Opt., 17, 1978, pp. 1748-55

[7] Dhir, S.K. and Sikora, J.P., "An improved method for obtaining the general displacement field from a holographic interferogram", Exp. Mech., 12, 1972, pp. 323-327

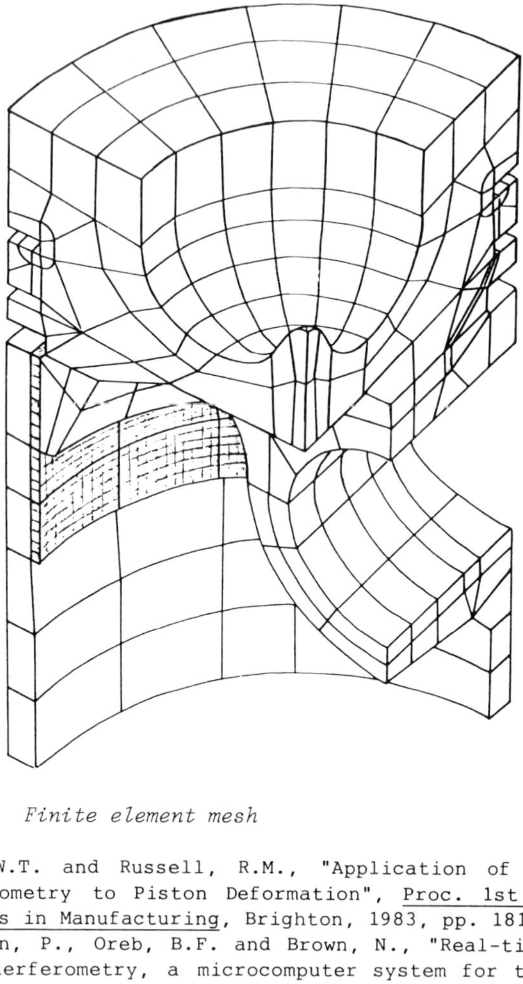

Figure 19.5: Finite element mesh

[8] Evans, W.T. and Russell, R.M., "Application of Holographic Interferometry to Piston Deformation", Proc. 1st Int. Conf. on Lasers in Manufacturing, Brighton, 1983, pp. 181-192

[9] Hariharan, P., Oreb, B.F. and Brown, N., "Real-time Holographic Interferometry, a microcomputer system for the measurement of vector displacements", Appl. Opt., 22, 1983, pp. 876-880

[10] Evans, W.T., Russell, R.M. and White, A., "Application of Holographic Interferometry to cylinder deformation", J. Strain Analysis, 1987, 22, 2, pp. 63-74

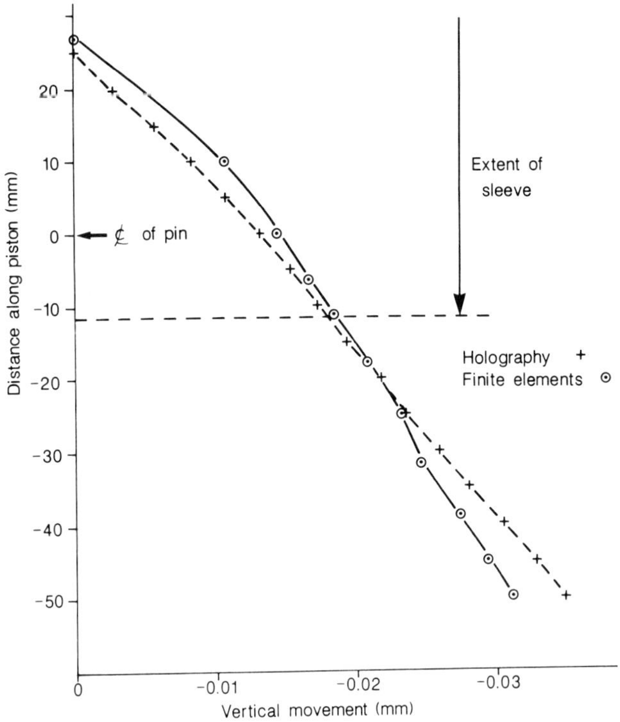

Figure 19.6: *Comparison between finite element and holographic results*

20 A workmanlike Moiré interferometer: conceptual and design considerations, operation, application, variations and limitations

J.McKelvie, C.A.Walker & P.M.MacKenzie

INTRODUCTION

The 1960's saw much development of the application of the Moiré phenomenon for the purpose of strain measurement, but the rate of development then appeared to slow down through the early 1970's. This de-emphasis was most probably due to the advent of coherent-optical techniques such as holographic interferometry and speckle methods, which on the face of it overcame at a stroke the two perceived disadvantages of Moiré, namely the necessity to treat the specimen surface (to form a grating or grid), and the displacement sensitivity obtainable (of the order of, in practice, 25 microns - as opposed to 0.5 microns typical of the new techniques).

The introduction of coherent-optical Moiré interferometry [e.g. 3,4] removed the discrepancy in respect of sensitivity but the necessity for surface preparation continues to be regarded - quite understandably - as a rather bothersome business, in comparison with the techniques for which the only treatment required on the surface is to illuminate it.

It transpires, however, that this self-same irksome process bestows a number of cardinal benefits which affect, with considerable impact, many important practical applications. This chapter explains these benefits and describes an embodiment of Moiré interferometry (and relevant applications thereof) in which advantage is taken of the beneficial aspects.

THE BENEFITS OF MOIRÉ-TYPE METHODS

Regular vs. random diffraction

Requirement of in-plane measurements

It is obvious upon minimal consideration that a mirror-like surface can provide no information about in-plane strains or displacements - at least in the case of observations in visible light. The surface must have "marks" of some kind on it so that we can make measurements before and after deformation. That is, the surface must diffract light. (Obviously we could use light of a wavelength sufficiently short to be diffracted by whatever fine-scale markings there inevitably are, or, trivially, we are free to use the specimen edges). It is of interest that this restriction does not apply to out-of-plane deformations, and by corollary to any cases in which in-plane can be directly related to out-of-plane - but these are exceptional. The optical measurement of in-plane deformation depends in essence therefore on the comparison of the "before" and "after" diffraction patterns of the surface.

Regular and irregular compared

The essence of the surface treatment applied for the Moiré method is the imposition of regularity on the surface. This alters the diffraction pattern so that it becomes structured, with much of the light (ideally all of it) being propagated in preferred directions, with smooth wave-fronts - as opposed to the diffuse propagation with random phase variation associated with an irregular scattering surface. Many surfaces, e.g., on machined components, do have a degree of regularity - and therefore of directionality - but the random element is usually strongly dominant on as-formed surfaces.

The essential differences are illustrated in Figure 20.1, which illustrates a small section of one regular and one irregular specimen each subsequently subjected to a strain condition consisting of an overall but non-uniform compression. The important point is that the diffracted wavefront, P_R from the undistorted regular specimen is flat. This means that when the distorted wavefront P_{RD} is superimposed on it, the positional registration is of no concern, so that a smoothly changing intensity will always be observed in the interference pattern. However, with the irregular object, the two wavefronts P_S , P_{SD} , run out of correlation towards one end, giving an interference pattern that itself tends to develop a random intensity. Furthermore, it is clear that rigid-body motion parallel to the surface will also cause decorrelation in the irregular case but will have no effect in the regular one. These factors are the basis of three important benefits of the Moiré technique.

Benefits of positional insensitivity

These are almost self-evident in the light of the last section and consist of:

"Off-the-table" operation

It is an important practical aspect that with a conventional Moiré, the in-plane registration of the "master" is of no consequence; (This is an exemplification of the point made previously). However, any out-of-plane movement causes fringes to appear due to perspective, and can be of considerable moment unless appropriate precautions are taken, e.g. tele-centric viewing. With Moiré interferometry using two colliminated beams (Figure 20.2) and using the convenient conceptualisation of the method in which a "virtual master grating" is imagined to be created in space,

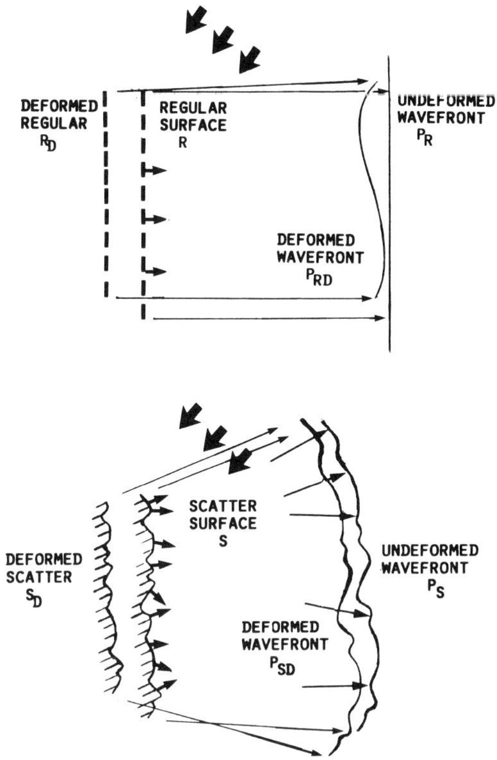

Figure 20.1: Diffraction at regular and random surfaces

we see that we are concerned with a volume of master grating envisioned as walls extending in the Z-direction, and it is obvious that the specimen's Z-position within the volume if irrelevant. This attribute is confined of course to the collimated set-up, which may itself be regarded as a restriction, akin in a way to tele-centricity. Thus we see that there are no spurious fringes introduced by positioning errors in the x, y or z directions. Moreover, it is equally obvious from Figure 20.2 that no spurious fringes are produced by rotation about the x-axis. Rotation about the z-axis introduces merely the so-called "rotational fringes", which for typical accidental misalignments do not affect the strain information, and we are left with the fact that the only misalignment which will cause undesirable "spurious strain" fringes is a rotation about the y-axis. It is shown [5] by analysis that the strain error associated with a rotation $\Delta\alpha$ in this degree of freedom is $\Delta\alpha^2/2$. In normal circumstances this is no great problem, being only 25 microstrain for 0.5° rotation - a substantial angle in terms of a great number of test situations. Of course if there is significant differential

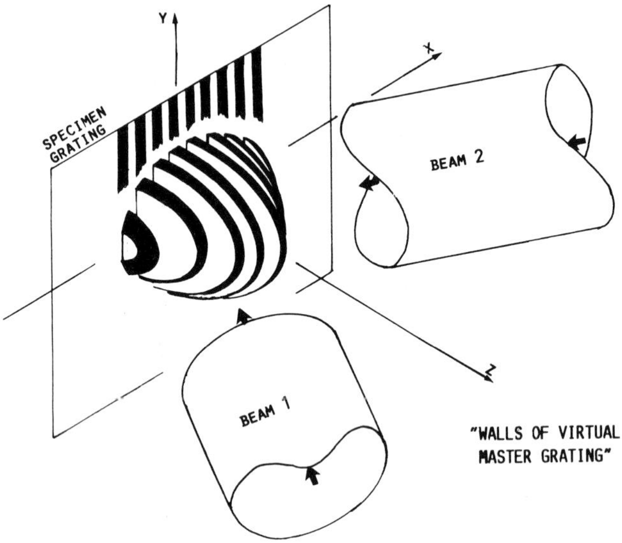

Figure 20.2: *Classical Moiré interferometry and concept of "virtual grating"*

rotation, for example crack-tip "dimpling", then a knowledge of the out-of-plane profile is necessary for precision work.

A crucial consequence of these characteristics is that the practitioner has liberated positional constraint (with the exception of a modicum of care in respect of one rotation), and is therefore free to operate in common work-place conditions - as opposed to being restricted to the optical table as is the case with other interferometric techniques. (The case of double pulse holography is not of relevance in the context of strain measurement. A corollary is that long-term monitoring in service becomes feasible.

Real-time-viewing

The discussion of "off-the-table" operation must not be construed as implying that the fringe pattern observed off-the-table is stationary. It is not possible - without some "bolting together" arrangement - to maintain the registration of the specimen grating and the "master" grating to the submicron tolerance that would be necessary to maintain fringe stability. In fact, "normal" vibrations induce translations and in-plane rotations which cause, respectively, phase changes of the Moiré pattern and the introduction of varying amounts of rotational fringes. However, it is axiomatic that such disturbances cannot be simultaneously large in both amplitude and frequency, without involving huge forces. The outcome is that in the vast bulk of applications the fringes, although continuously moving, are quite visible and this is in practice of great benefit in most situations.

This is in direct contrast to, for example, speckle photography, which, although to a degree tolerant of positional de-registration, is in its very principle a "double exposure" technique necessitating "blind" operational assessment. A corollary in this instance is the possibility of continuous recording of a test.

Immunity to large movements

In testing sizeable specimens under realistic loads, large whole-body displacements (e.g. several millimetres) can occur. These are of no particular consequence in the Moiré method.

Suppression of speckle effects

One of the most important aspects of the Moiré interferometry is the very low speckle level which is obtained. This results, again, from the directional nature of the diffracted light. A randomly-scattering wavefront will occupy virtually the entire hemisphere available to it. An infinitely large perfect lens would collect all the light and reconstruct a perfect image. But a real lens captures only a fraction, over a limited scatter angle, and this gives rise to the phenomenon of speckle, which is in interferometry generally regarded as a nuisance because it limits the quality of the information in a fringe pattern and puts a restriction on the maximum fringe density - and therefore on both the highest strain level and the smallest gauge length - attainable.

On the other hand, it is perfectly feasible to have a lens large enough to capture completely a directionally-diffracted wave-front of some significant dimension (say 150mm) and to produce interference patterns which are almost speckle-free. This enables very high fringe densities to be observed, yielding a concomitant wealth of detail. Of course, there is always some element of scattered light in addition to the directional one, and the relative level will vary according to the grid-making process. For example, chemical etching of grids in metals gives a strong scatter, whereas the replication technique (Figure 20.3) favoured by the authors, throws the bulk of the light into the diffracted orders.

Toleration of plastic deformation

One of the most useful aspects of the Moiré approach is that the imposed regularity can persist - and even predominate - in the presence of considerable plastic flow. Under this condition, of course, other methods simply cannot function since the "before" and "after" surfaces have almost no mutual correlation whatsoever. With the Moiré method there is, it is true, a gradual degradation of the plane wavefront as plasticity progresses, but entirely satisfactory fringe patterns can be obtained in the presence of strains of several percent. The authors have recorded strains beyond 10% in titanium tube (see below) with a system having a sensitivity of 1 micron.

Efficient use of light

Because the light is concentrated into a few preferred directions

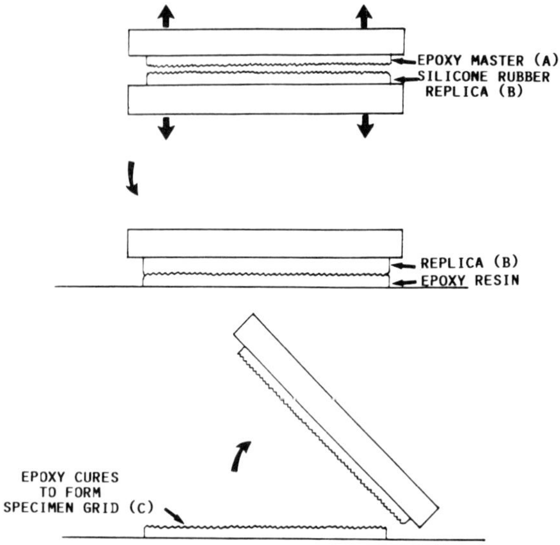

Figure 20.3: Grid replication process

and indeed with modern grating technology can be optimised as desired, it becomes feasible to utilise low powered lasers. Of course, as mentioned above, the fringes (once we move off the optical table), are continuously on the move and we require a fast exposure to "freeze" them, typically 1/500 sec. Using efficient gratings the authors have routinely photographed fringe patterns over an area 50mm dia. using a 5mW He-Ne laser. This aspect is of some significance when considering a portable instrument.

DESIGN OF A PORTABLE MOIRE INTERFEROMETER

Conceptual

The basis of the design is a specification of an instrument which can measure three components of in-plane strain with a sensitivity comparable to that of holography, covers the maximum reasonable aperture obtainable in a completely self-contained, reasonably portable unit, which has sufficient mechanical stability to tolerate the off-the-table environment, and is convenient to align and operate. In practice this resulted in the following system parameters: grid pitch 1 micron effective, aperture 50mm dia. (the resulting configuration represented a compromise among several criteria - size of instrument, stand-off distance, sensitivity).

The decision to opt for three-component sensing was imposed by the requirement to operate off-the-table. For whereas it is entirely possible to determine the shear strain by combining U and V-displacement information, that particular process demands that any whole-body rotation occuring between those two recordings be precisely known - and the workaday environment precludes this. It conveniently happens that an orthogonal crossed grid can be interrogated at 45° to yield a third strain component and the strain tensor can then be determined in the usual "rosette" manner.

A three beam system

The usual configuration for the measurement of one strain component is two symmetrically disposed beams as in Figure 20.2. The corollary is that six beams would be required in order to measure three components. This was deemed undesirable on two counts - the optical complexity, and the "waste" of two-thirds of the light (since any one pair would employ only one-third, in the absence of some switching system - which in turn would have involved even more complexity).

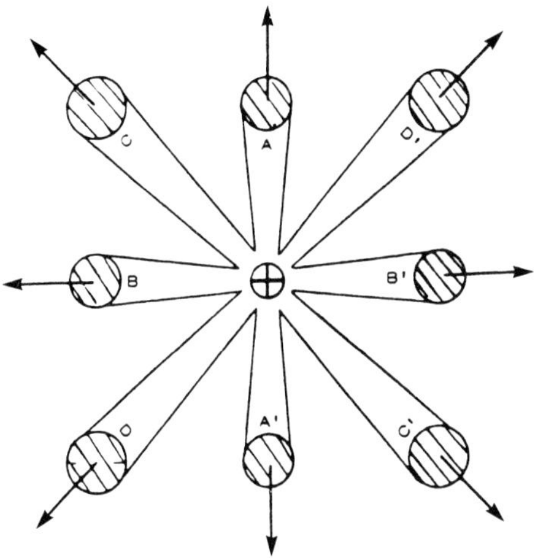

Figure 20.4: Diffraction by an orthogonal grid

The layout chosen is based upon the fact that in Figure 20.2 a rotation about the x-axis has no effect. Figure 20.4 illustrates the diffraction pattern of a beam impinging normally on an orthogonal grid. By the principle of reversibility of optical paths, we may direct beams back down, for example, beams D' and C', and make observations along the normal; it transpires that the desired interference pattern (corresponding to input beams AA') appears in this channel. The design therefore utilises three such beams, which taken in pairs, yield the three desired components with only one-third of the light being "wasted". The arrangement is illustrated in Figures 20.5 and 20.6.

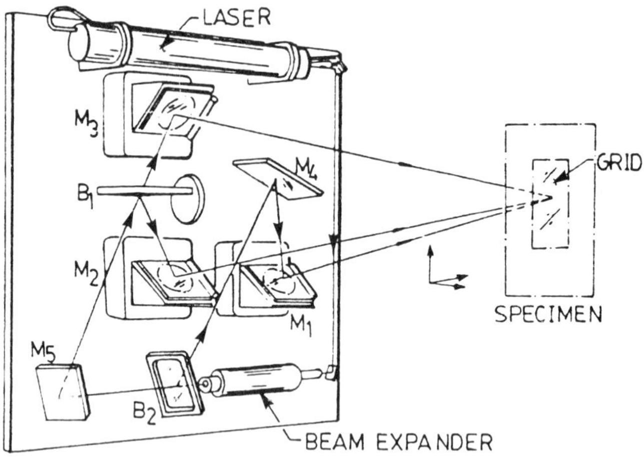

Figure 20.5: Schematic of interferometer

Figure 20.6: Actual interferometer

Polarisation control

The control of polarisation is necessary to permit interference and when all incident reflected beams are co-planar then no problems arise. However, in the geometry illustrated, it is axiomatic that non-co-planar reflections occur at the final mirrors and maintaining compatibility in polarisation is not straightforward. It turns out that it is rather awkward to design for three beams of the desired geometry all with mutually ideal polarisation planes. A good compromise was proposed that the intersections of the three planes of polarisation with the specimen surface should all be parallel. An arrangement was devised which produced this condition (Figure 20.7). With polarisation initially parallel to the base-plate, then any mirrors whose intersections

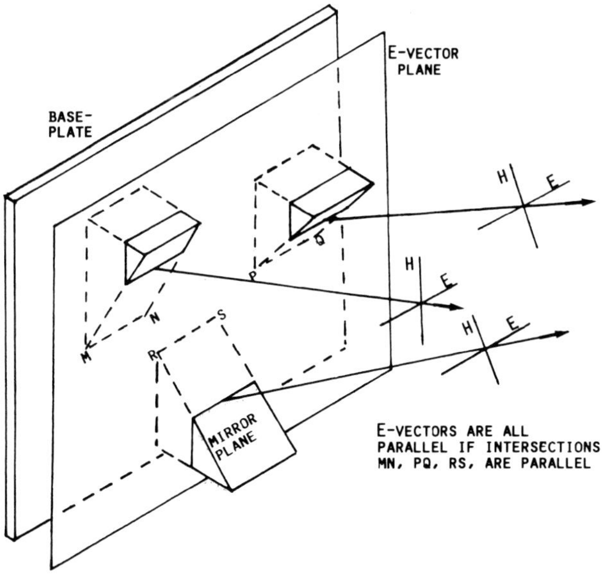

Figure 20.7: *Arrangement for parallelism of polarisation planes in three emerging beams*

with the main base-plate are parallel will have reflected beams whose planes of polarisation are also parallel.

This arrangement, though not perfect, produces adequate interference quality. The polarisation rotations which occur after diffraction at the specimen surface were studied, and for the rather small angles of incidence (17°) in the chosen geometry, little further rotation occurred and a very acceptable fringe quality is obtained. The extrapolation of the above method to cases where the sensitivity involves much larger angles of incidence would, however, not be particularly valid. Dielectric mirrors are used throughout to avoid ellipticity.

Alignment system

The interferometer is hung on a yoke which is attached to the centre of the back plate and which has three rotational degrees of freedom. A 100mm simple lens at the front collects the desired diffracted orders and focuses them in a convenient plane at the rear of the yoke. The focussed spots immediately indicate the relative alignment of interferometer and specimen, and it is literally only two or three minutes work to align the system so that all three dots coincide in the optical axis.

Mismatch facility

One of the most useful aspects of the Moiré method is the use of mismatch. Variable mismatch facility is built in by mounting the final mirrors on plates which are spring-retained against three ball-ended supports, of which two are screw-adjustable and one is a barrel micrometer. This latter permits pitch mismatch to be introduced at will. Rotational mismatch is easily introduced by the in-plane rotation mechanism of the yoke, which is worm-gear driven.

Spatial filtering

The practice of spatial filtering in order to "clean-up" the beam has been dispensed with. Experience indicated that the sensitivity of this type of filter to small amounts of beam "wander" was incompatible with the intended working environment. In practice, the resulting imperfections have proved to be insignificant.

OPERATION

The instrument is surprisingly straightforward to operate. Alignment to a specimen, as mentioned above, is a matter of a few minutes. Usually, for convenience, the yoke is mounted on a wheeled tripod with a motorised vertical adjustment, so that x-y-z positioning is also readily accomplished. The central mounting of the yoke has proved to be beneficial in respect of mechanical stability; operated in a "table-mounted" mode there is a tendency for the back-plate to bend a little, leading to the possibility of error. The centrally hung arrangement does not exhibit this effect. The stand-off distance is some 250mm which is convenient and work has been carried out via a mirror to project the beams into awkward corners.

The switching from x to y to 45° components is achieved by sequentially shuttering each beam in turn. (The shutters are the only moving parts in the instrument).

Video recording has been used. With the commonplace 25fps raster, the moving fringes are quite visible even though individual frames tend to show poor fringe visibility due to the rather long exposure time. With modern fast-exposure cameras this should not be a problem. Using a grid deposition in techniques described below, light levels are such that exposures in the range 1/250-1/500 sec. are adequate using Kodak Tri-X Pan film developed to an effective ASA 1600 at a demagnification of 1:2. With ordinary colour video cameras there is a tendency to saturate at this magnification.

GRID TECHNOLOGY

The grid deposition process is based upon replication of phase-type grids as is the common practice for moire interferometry [5,6]. All specimen grids are descended from one master grid manufactured by the National Physical Laboratory (UK) to our bespoke specification, with particularly good diffraction into the (1,1), (1-1), (-1,1) and (-1-1) orders. Many hundreds of specimen grids have been produced in sizes up to 95mm square and the process is quite routine.

No aluminising of the grids is used except where particularly short exposure times may be required (see example below). Instead, the resin is dyed green to prevent back-scatter from reflective substrates. Normally, no surface treatment other than thorough cleaning is necessary, the resin usually accommodating any natural surface roughness. In cases where the surface is particularly rough or undulating, a "flattening coat" of resin is applied first. Curved surfaces are dealt with in similar fashion, i.e. they are converted to a "flat" over a suitably restricted arc.

There are cases when silicone rubber specimen grid is preferable to epoxy (see below). There is a loss of light in these instances due to the lower reflectivity of the rubber.

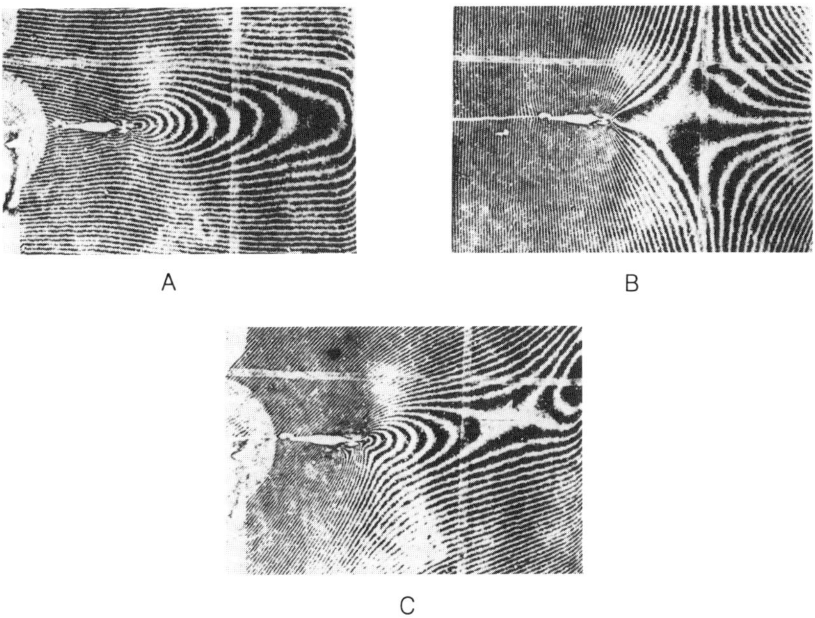

Figure 20.8a, b, c: *X, Y, 45° component on a cracked CT specimen*

APPLICATIONS

Fracture mechanics

The system has been used to measure the J-Integral in steel [7] and in aluminium and titanium [8]. Figures 20.8a, 8b and 8c show typical contours which are used to evaluate J directly by integration round suitable contours outside the plastic region. Validation against J measured simultaneously by conventional methods on standard specimens is shown in Figure 20.9.

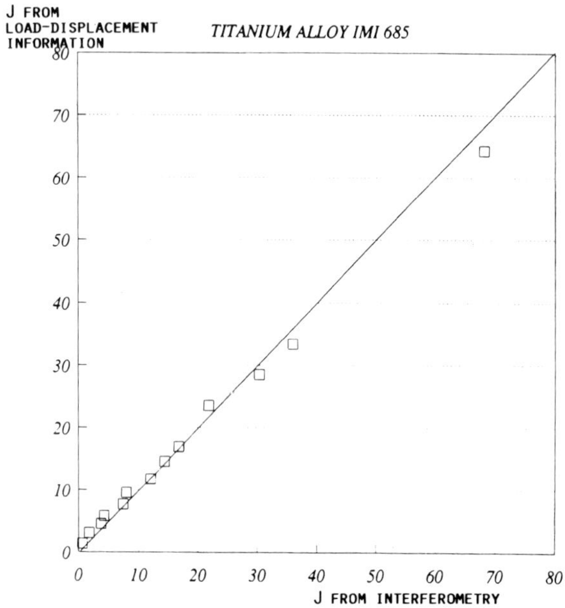

Figure 20.9: Comparison of J obtained interferometrically and conventionally

A problem has been encountered with very tough steel in the standard deeply-cracked 3-point bend geometry. At high loads, plasticity encroaches towards the crack from the central load point and the whole section may yield before initiation. The inferring of stresses in these conditions is problematical. However, it is a unique feature of this method of measuring J that specimen geometry and crack-length are of no concern, and the use of alternative geometries becomes possible. Indeed, application on actual structures is entirely feasible.

Plasticity studies

Figure 20.10a shows a Moiré interferogram of the displacement field arising from the interference fitting of a bush into an aluminium lug. An alternative process which gives rise to extensive plasticity gives the result shown as Figure 20.10b. By way of comparison, Figure 20.10c is a holographic interferogram (of a slightly different lug geometry) corresponding to Figure 20.10b, and Figure 20.10d is another holographic interferogram describing the elastic relaxation when a bush is removed (having been inserted according to the method of Figure 20.10a.

This series of interferograms illustrates succinctly the appropriateness of holographic interferometry in the one case and the relevance of the Moiré technique in the presence of plastic flow. The authors are indebted to Rolls Royce Ltd. and to the UK National Physical Laboratory for permission to use Figures 10a, 10b, 10c and 10d in the present context].

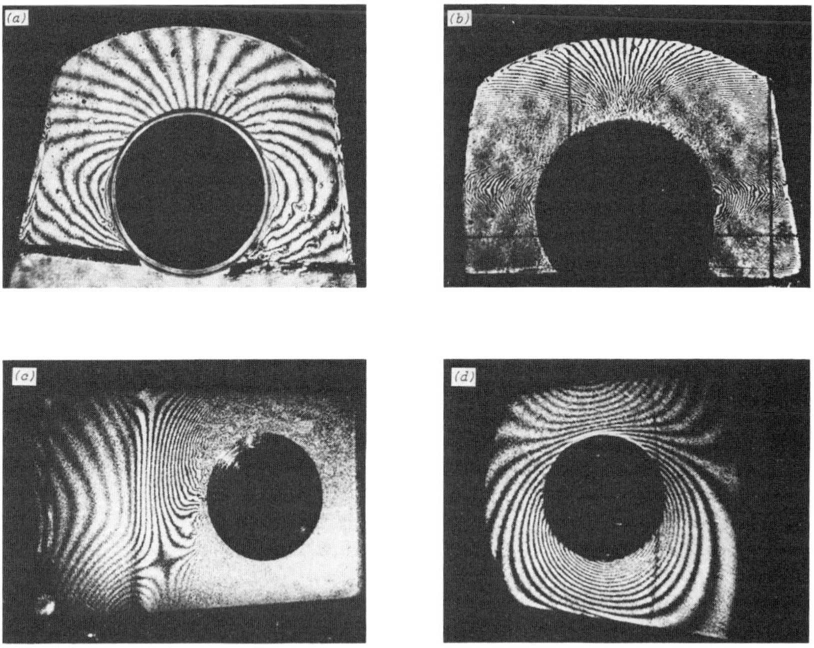

Figure 20.10: (a) *Moiré interferogram of interference fitted bush (courtesy Rolls Royce plc)*
(b) *Similar to 10a with much greater interference (courtesy Rolls Royce plc) [Fringes are visible over the whole field in the original]*
(c) *Holographic interferogram equivalent to 10b (courtesy Rolls Royce and NPL)*
(d) *Holographic interferogram resulting from bush removal (courtesy Rolls Royce and NPL)*

Figure 20.11 shows the field around a crack tip. There is a fringe density in the region of 30 per mm (750 per inch), which demonstrates the advantage of the relative absence of speckle. (The fringe is an enlargement from a 50mm dia. area).

Figures 20.12a, b, and c show a 16mm dia. titanium tube and the interferograms obtained from it at some 8% and 10% strain. The eventual break-up of the fringes (due to substrate surface changes) is obvious. Mismatch was introduced progressively through the test to keep the fringe density reasonably low. The fringe quality is poor at this stage - but still usable; the sensitivity remains one micron.

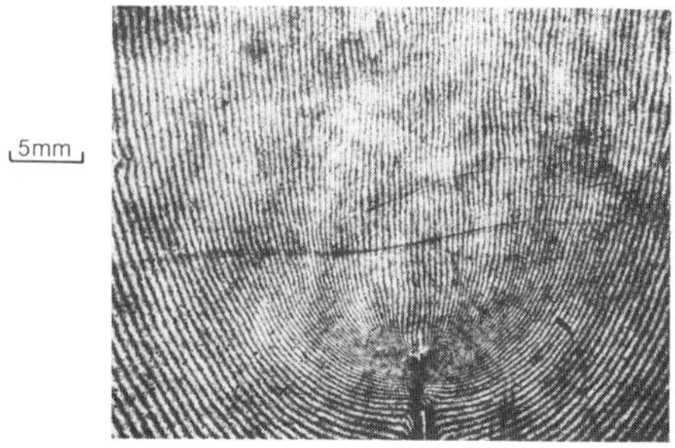

Figure 20.11: *Example of high fringe densities obtainable*

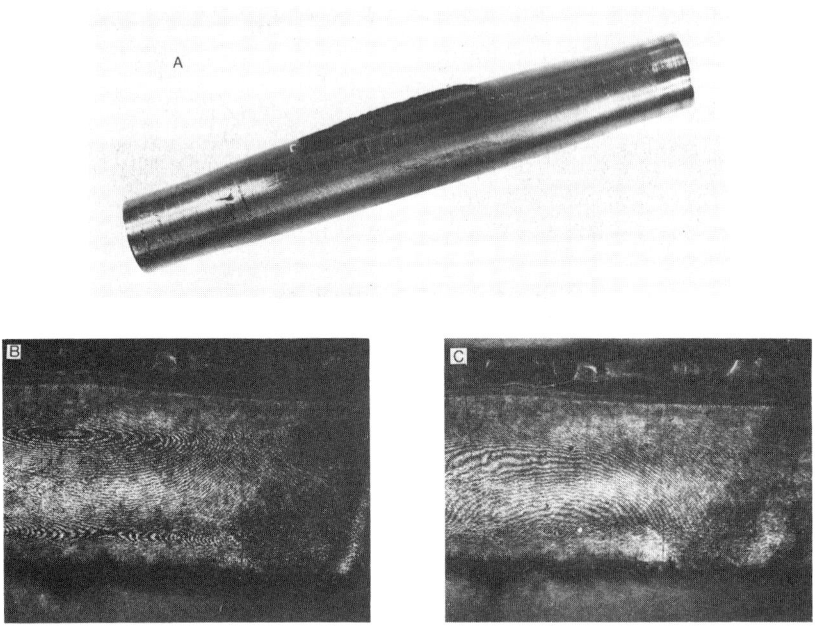

Figure 20.12: *(a) 16mm dia. titanium tube with gridded "flat"*
(b) Moiré interferometry fringes at 8% strain
(c) Moiré interferometry fringes at 10% strain

Dynamic studies

A titanium tube-plate was studied while tubes were being rolled in. The rolling of tubes is a fairly violent process, and in order to freeze the fringes, an exposure of 1/2000 sec. was used. To facilitate this, the grid was vacuum-aluminised after being applied. A motorised camera taking some six frames per second was employed, and thousands of frames were successfully produced during the rolling of a series of tubes [9]. Figure 20.13 shows several consecutive frames. The possibility for high-speed cine or video is obvious.

Residual stress

The method has been used for residual stress measurement by the hole-drilling technique, and the results are consistent with parallel measurements made by the standard three-gauge method. There are some advantages in using moiré for this purpose in certain situations.

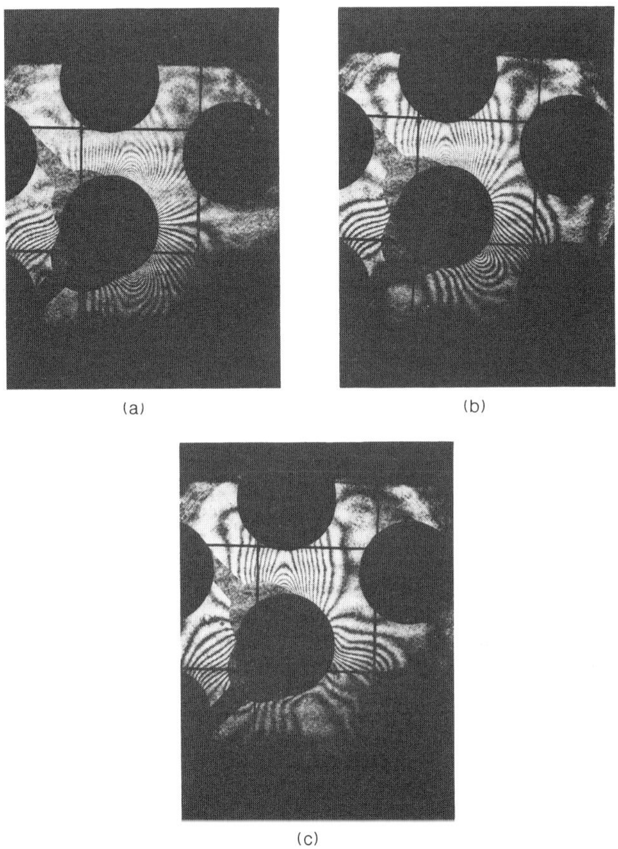

Figure 20.13: *Sequence of interferograms during "live" rolling of titanium tubes*

VARIATIONS

Silicone rubber grid

Certain circumstances invoke a preference for the use of silicone rubber - temperature considerations and very high strains (particularly compression). For instance, the bushing examples quoted above utilised silicone. The titanium tube with 10% extension was also performed with the alternative material. In severe compression there is a tendency for epoxy to debond, but with silicone, due to its much lower ratio of stiffness/adhesive-strength, this has never been observed.

Simultaneous three-beam recording

It is entirely practicable to capture the information about all components in one exposure by simultaneous use of all three beams. This is analogous to the "grid analyser method" [10], with the difference that there is three-beam interference, leading to some mutual modulation of the different sets of fringes. Figure 20.14 shows the lug of Figure 20.10a with a three-beam exposure.

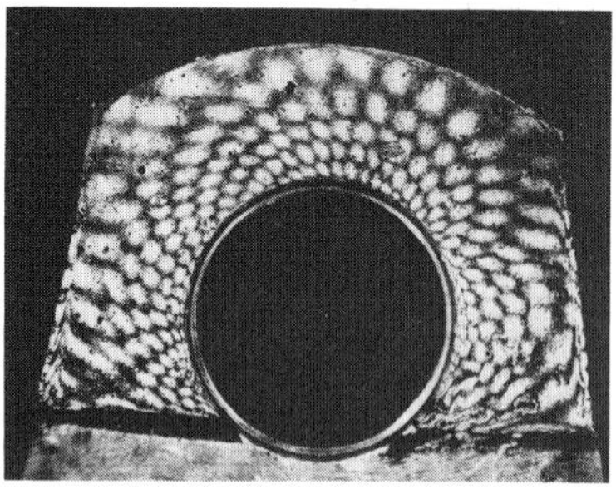

Figure 20.14: Three-beam exposure, otherwise as Figure 10a

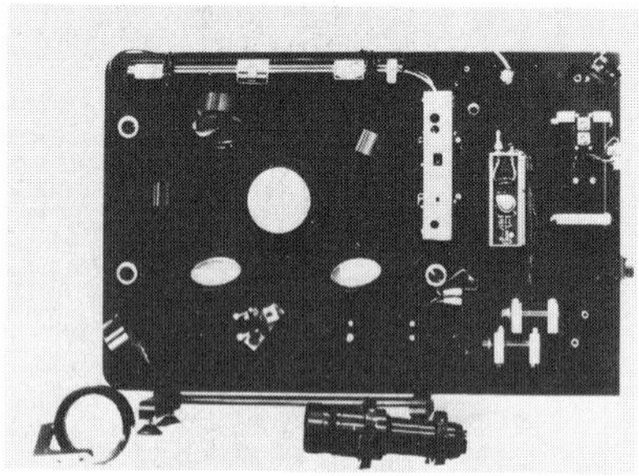

Figure 20.15: Interferometer with heterodyne/"bolt-on"

This variation would be useful in dynamic situations where all strain components are required. (It also increases the light available by 50%).

Heterodyning

The interferometer has been successfully modified to operate in a heterodyne mode on a trial basis. The conversion is quite readily carried out - the extra components required being permanently mounted on a "bolt-on" plate (Figure 20.15).

Only one component has been so converted, and trials [11] have indicated that the method has promise. The potential for rapid, precise, low-gauge length whole-field, off-the-table strain measurement is probably not equalled in any other proposed device.

LIMITATIONS

Curved surfaces

Practically all strain measurement systems are awkward to use on curved surfaces, for one reason or another. It has been the authors' practice to create a local "flat". This is succussful but it puts a considerable limit on the arc that can be looked at if the curvature is pronounced.

Restricted field

It is a beauty of holography and speckle photography that they can use expanded beams for illumination, and thus cover large areas. With Moiré interferometry, collimated light is used and so the field is restricted by the expense (and unwieldiness) of large collimating systems.

Vignetting

The very feature which leads to the benefits mentioned above, namely the directional nature of the diffracted light, also has one drawback. If the specimen rotates differentially then the

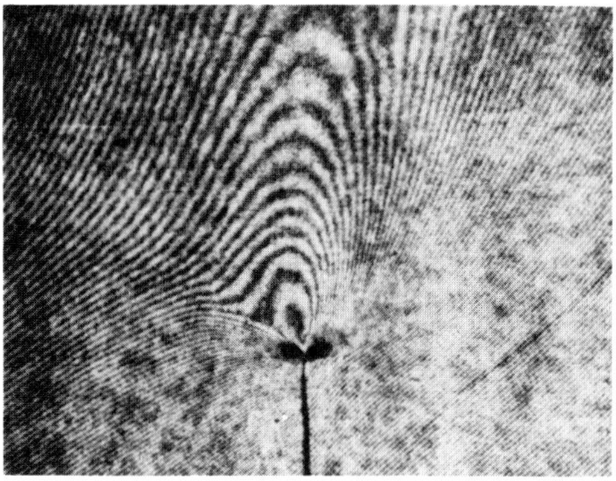

Figure 20.16: Vignetting at a crack-tip dimple

various areas will diffract in different directions. It may not then be possible to accommodate all the diffracted light in the one recording optical system, and vignetting occurs. Figure 20.16 illustrates the effect from heavy "dimpling" ahead of the crack. In such a case it is possible, by suitably rotating the interferometer, to view separately the two faces of the dimple. The aperture of the imaging system should be kept as wide as possible to avoid the vignetting effect.

CONCLUSIONS

A Moiré interferometer suitable for everyday use has been described and the particular advantages and the basic principles underlying them have been explained.

Moiré interferometry is a technique having attributes which render it especially useful in everyday work, and various applications illustrating this have been reported. Some limitations have been identified and new developments have been initiated which offer great potential in the field of strain measurement.

REFERENCES

[1] Durelli, A.J. and Parks, V.J., "Moiré Analysis of Strain", Prentice-Hall, Englewood Cliffs, N.J. (1970)

[2] Theocaris, P.S., "Moiré Fringes in Strain Analysis", Pergamon, Oxford (1969)

[3] McDonach, A., McKelvie, J. and Walker, C.A., "Stress Analysis of Fibrous Composites using Moire Interferometry", Opt. Lasers Eng., 1 (1980), pp. 85-105

[4] Post, D., "Optical Interference for Deformation Measurements] Classical, Holographic and Moiré Interferometry", in Mechanics of Non-destructive Testing, Stinchomb, W.W., ed., Plenum Press, New York (1980), pp. 1-53

[5] McDonach, A., McKelvie, J., MacKenzie, P. and Walker, C.A., "Improved Moiré Interferometry and Applications in Fracture Mechanics, Residual Stress and Damaged Composites", Exp. Techniques, 7 (6) (1983), pp.20-24

[6] Post, D., "Moiré Interferometry" in Handbook on Experimental Mechanics, Prentice-Hall, Englewood Cliffs, N.J. (1987)

[7] Gray, T.G.F., McKelvie, J., MacKenzie, P. and Walker, C.A., "Interferometric Measurement of J for Arbitrary Geometry and Loading", Int. J. Fracture, 24 (1984), R109-R114

[8] MacKenzie, P., McKelvie, J. and Walker, C.A., "Evaluation of the J-Integral using a High-Sensitivity Moiré Technique", Proc. 1986 SEM Spring Conference on Experimental Mechanics, New Orleans (1986), pp. 773-780

[9] Walker, C.A., McDonach, A., MacKenzie, P. and McKelvie, J., "Dynamic Moiré Measurement of Strains Induced in a Tubeplate During the Rolling of a Series of Tubes", Experimental Mechanics, (1985), pp. 25, 1, 1-5

[10]Post, D., "The Moiré Grid Analyser Method for Strain Analysis" Proc. SESA, 22 (1965), pp. 358-377

[11]Walker, C.A., MacKenzie, P., and McKelvie, J., "Heterodyne Moiré Interferometry - Some Experience of the Development of a Rapid-Measurement System", Proc. 1986 Spring Conference on Experimental Mechanics, New Orleans (1986), pp. 695-699